Advanced Problems in Mechanics

Holm Altenbach · Vitaly Kuzkin
Editors

Advanced Problems in Mechanics

Springer

Editors
Holm Altenbach
Institute of Materials
Technologies and Mechanics
Otto-von-Guericke-University Magdeburg
Magdeburg, Germany

Vitaly Kuzkin
Institute for Problems in Mechanical
Engineering
Russian Academy of Sciences
St. Petersburg, Russia

ISBN 978-3-031-80102-0 ISBN 978-3-031-80103-7 (eBook)
https://doi.org/10.1007/978-3-031-80103-7

This Springer imprint is published by the registered company Springer Nature Switzerland AG
The registered company address is: Gewerbestrasse 11, 6330 Cham, Switzerland

Preface

The book, prepared on some lectures of the 50th Anniversary of the International Summer School-Conference "Advanced Problems in Mechanics" (June 2022, St. Petersburg, Russia), covers problems not limited to mechanical engineering, but includes almost all promising problems in mechanics, which are reflected in the title of the conference. The focus is on problems at the interface between mechanics and other fields, such as micro- and nanomechanics, materials science, solid state physics, molecular physics, astrophysics, and many others. One of the main goals of the conference is to transfer scientific experience from renowned scientists to their young colleagues.

The first Conference was organized by the outstanding scientist Jakow Gilelewitsch Panowko (1913–2002) and his colleagues in 1971 and focused on the nonlinear oscillations of mechanical systems with a finite number of degrees of freedom. The conference existed in this form until 1985. In the period from 1986 until 1993 the conference was organized in different forms. In 1994, the Institute for Problems in Mechanical Engineering of the Russian Academy of Sciences revived the tradition of holding an annual school-conference. The new impetus to the school-conference was given by well-known mechanicians Dmitry A. Indeitsev (1948–2023), Nikita F. Morozov (1932–2024), Vladimir A. Palmov (1934–2018), Pavel A. Zhilin (1942–2005), and their younger colleagues. Participants could enjoy a stimulating atmosphere and engage in lively debates and friendly, passionate scientific discussions. The school-conference acquired the status of an international one, its scope was significantly expanded and today covers not only all areas of mechanics, but also interdisciplinary issues. In recent years the school-conference was co-chaired by Dmitry A. Indeitsev and Anton M. Krivtsov.

The Scientific Committee of the 2022 conference was formed by D. A. Indeitsev (IPME RAS, Peter the Great St. Petersburg Polytechnic University, Russia)—Co-chairman (he passed away in December 2023); A. M. Krivtsov (Peter the Great St. Petersburg Polytechnic University, IPME RAS, Russia)—Co-chairman, P. A. Dyatlova (Peter the Great St. Petersburg Polytechnic University, IPME RAS, Russia)—Scientific secretary; H. Altenbach, Otto von Guericke University Magdeburg, Germany; M. B. Babenkov, IPME RAS, Peter the Great St. Petersburg Polytechnic

University, Russia; V. A. Babeshko, Southern Scientific Center RAS, Rostov-on-Don, Russia; A. K. Belyaev, IPME RAS, Peter the Great St. Petersburg Polytechnic University, Russia; I. E. Berinskii, Tel Aviv University, Israel; V. A. Bratov, IPME RAS, St. Petersburg, Russia; A. A. Burenin, Institute of Metallurgy and Mechanical Engineering Far-Eastern Branch of RAS, Komsomolsk-na-Amure, Russia; A. V. Cherkaev, University of Utah, Salt Lake City, USA; F. Dell'Isola, Università di Roma La Sapienza and MEMOCS centre, Italy; V. A. Eremeyev, Gdansk University of Technology, Poland; V. I. Erofeev, Mechanical Engineering Research Institute of RAS or MERI RAS, Russia; A. B. Freidin, IPME RAS, Peter the Great St. Petersburg Polytechnic University, Russia; M. E. Frolov, Peter the Great St. Petersburg Polytechnic University, Russia; S. N. Gavrilov, IPME RAS, St. Petersburg, Russia; I. G. Goryacheva, Institute for Problems in Mechanics RAS, Moscow, Russia; E. F. Grekova, IPME RAS, St. Petersburg, Russia; University of Seville, Spain; N. Gupta, Indian Institute of Technology Delhi, India; M. A. Guzev, Institute of Applied Mathematics of Far-Eastern Branch of RAS, Russia; H. E. Huppert, University of Cambridge, UK; H. Irschik, Johannes Kepler University of Linz, Austria; M. L. Kachanov, Tufts University, Medford, USA; B. L. Karihaloo, Cardiff University, UK; D. A. Koroteev, Center for Hydrocarbon Recovery, Scoltech (Digital Petroleum), Russia; S. V. Kuznetsov, IPMech RAS, Russia; V. A. Kuzkin, Peter the Great St. Petersburg Polytechnic University, IPME RAS, Russia; V. A. Levin, Lomonosov Moscow State University, Russia; A. M. Linkov, IPME RAS, Russia; Rzeszow University of Technology, Poland; I. I. Lipatov, Moscow Institute of Physics and Technology, Russia; O. S. Loboda, Peter the Great St. Petersburg Polytechnic University, IPME RAS, Russia; E. V. Lomakin, Lomonosov Moscow State University, Russia; G. S. Mishuris, Aberystwyth University, UK; N. F. Morozov, St. Petersburg State University, IPME RAS, Russia (he passed away in July 2024); U. Nackenhorst, Leibniz University of Hannover, Germany; E. E. Pavlovskaia, University of Aberdeen, UK; S. V. Petinov, IPME RAS, Peter the Great St. Petersburg Polytechnic University, Russia; Y. V. Petrov, St. Petersburg State University, IPME RAS, Russia; A. V. Porubov, IPME RAS, St. Petersburg, Russia; D. A. Prikazchikov, Keele University, UK; J.-N. Roux, Université Paris-Est, Laboratoire Navier, France; M. B. Rubin, Israel Institute of Technology, Haifa, Israel; A. I. Rudskoi, Peter the Great St. Petersburg Polytechnic University, Russia; S. Rudykh, University of Wisconsin-Madison, USA; S. H. Sargsyan, Gyumri State Pedagogical Institute, Armenia; V. V. Sergeev, Peter the Great St. Petersburg Polytechnic University, Russia; I. Sevostianov, New Mexico State University, USA (he passed away in July 2021); M. Simonov, Gazpromneft Science and Technology Center, Russia; A. Sokolov, TU Berlin, Germany; M. Wiercigroch, Aberdeen University, Scotland; H. A. Wu, University of Science and Technology of China, Chinese Academy of Sciences; P. Venkitanarayanan, Indian Institute of Technology, India; E. N. Vilchevskaya, IPME RAS, Peter the Great St. Petersburg Polytechnic University, Russia.

Presentations devoted to fundamental aspects, or widening the field of applications of mechanics, are invited. We are particularly keen to receive contributions that show new effects and phenomena or develop new mathematical models. The topics of the conference cover all fields of mechanics, including, but not restricted, to:

- complex media: micropolar theory, chemomechanics, acoustic metamaterials etc. (this section was dedicated to the memory of P. A. Zhilin)
- fluid mechanics
- nano-, micro- and mesomechanics
- phase transitions and nonlinear elasticity
- solids and structures
- mechanical and civil engineering applications
- nonlinear and multibody dynamics, chaos and vibration
- quantum nanomechanics
- plasticity
- heat transfer and wave motion
- reliability of technical systems.

In addition, the following minisymposia were organized:

- Biomechanics organized by Olga S. Loboda (Peter the Great St. Petersburg Polytechnic University, Russia) and Svetlana M. Bauer (St. Petersburg State University, Peter the Great St. Petersburg Polytechnic University, Russia).
- Nonlinear waves in continuous media organized by Vladimir I. Erofeev (Mechanical Engineering Research Institute of the Russian Academy of Sciences or MERI RAS, Nizhniy Novgorod, Russia) and Alexey V. Porubov (IPME RAS, St. Petersburg, Russia).
- Mathematical modeling in petroleum engineering organized by Kristina A. Plokhova (Peter the Great St. Petersburg Polytechnic University, Russia), Vitaly A. Kuzkin (Peter the Great St. Petersburg Polytechnic University; IPME RAS, Russia) and Galina Ayupova (Peter the Great St. Petersburg Polytechnic University, Russia).

The selected articles are presented below in alphabetical order.

Magdeburg, Germany
St. Petersburg, Russia
October 2024

Holm Altenbach
Vitaly Kuzkin
Ekaterina Saprykina

Contents

Chapter 1
Uniaxial High Strain Rate Tension Method Using Pulsed Magnetic Pressure

Dmitrii Alekseev, Dmitrii Zhukov, Sergey Krivosheev, Sergey Magazinov, and Eugeny Ostropiko

Abstract The mechanical properties of the materials depend on strain rate and there are various methods for testing mechanical properties with different strain rates. In this work we consider uniaxial high strain rate tension methods using pulsed magnetic pressure as a source of mechanical impact. In these methods pulse current generators (PCG) is discharged onto the magnetic system creating strong magnetic field. The geometry of magnetic system in the first method is a spiral coil near conductive disk connected to the sample, the other consist of flat closely spaced copper conductors that are placed inside gaps of the samples. Experiments were performed and finite element modelling was done. Pulsed magnetic field and pressure of the magnetic systems were calculated in the Comsol Multiphysics environment. Dynamic mechanical deformation of the samples is calculated in ANSYS AutoDyn software. Experiments were done on the oxygen-free high thermal conductivity (OFHC) copper and TiNi samples with strain rates in the range of 2000–7000 1/s. Registration of the deformation process using high speed camera or laser with photodetector were done. Modifications of the uniaxial tension using flat conductors, which allow to simplify analysis of the results and the preparation of the samples were made.

1.1 Introduction

It is well known that the mechanical properties of materials depend on the strain rate. For example, with increase in the strain rate, the flow stress of metals increases as well. There are various models which describe this effect, such as Johnson–Cook [1], Zerilli-Armstrong [2], Folansbee-Kocks [3], PTW [4], etc. Data on the parameters of these models is required for the analysis of structures under impact loads, for example in traffic accidents.

There are various methods for testing of the high strain rate mechanical properties of materials, such as Taylor impact [5], Hopkinson-Kolsky [6], Hauser [7], spall

D. Alekseev · D. Zhukov · S. Krivosheev · Sergey Magazinov (✉) · E. Ostropiko
Peter the Great St.Petersburg Polytechnic University (SPbPU), Saint Petersburg, Russia
e-mail: magazinov_sg@mail.ru

H. Altenbach and V. Kuzkin (eds.), *Advanced Problems in Mechanics*,
https://doi.org/10.1007/978-3-031-80103-7_1

fraction [8], etc. These methods have their advantages/disadvantages and require pulsed pressure impact, which can be created by compressed gas, detonation of explosive, collision with a moving body and so on.

Also, for these methods, the pressure impact for high strain rate testing materials can be created by using pulsed magnetic pressure. And magnetic pressure can be applied for samples in various loading schemes, such as tube expansion [9–11], three-point bending [12], Hopkinson-Kolsky [13, 14], spall fraction [15] and so on [16, 17]. Previously [18], we used magnetic pressure for the fracture test (mode I) of the copper samples, the results you can see in the Sect. 1.3.2. In this work we consider uniaxial tension methods using pulsed magnetic fields.

1.2 Methods of the Uniaxial Tension Using Pulsed Magnetic Pressure

To create pulsed magnetic pressure pulse current generator was used, see Sect. 1.2.1, which generate pulsed electric current in magnetic systems, such as the spiral coil near the conductive disk (see Sect. 1.2.2) and closely spaced flat conductors (see Sect. 1.2.3). Pulsed current carrying conductors generates magnetic field and creates pulsed magnetic volumetric Lorentz force $\overline{f}_l = \overline{J} \times \overline{B}$. This force is transferred to the test specimen and deforms it.

During testing, we can measure pulsed current using Rogowski coil and deformation of the sample using high speed camera or laser with photodetector. Knowing these experimental data, we can calculate magnetic pressure using finite element method (FEM) in the Comsol software and we can verify the plasticity models of the testing materials in the ANSYS AutoDyn FEM software (see Sect. 1.2.4).

1.2.1 Pulse Current Generator (PCG)

The main part of the PCG are capacitors C_{PCG}, which are discharged throw switch S_{PCG} to the magnetic system, see Fig. 1.1.

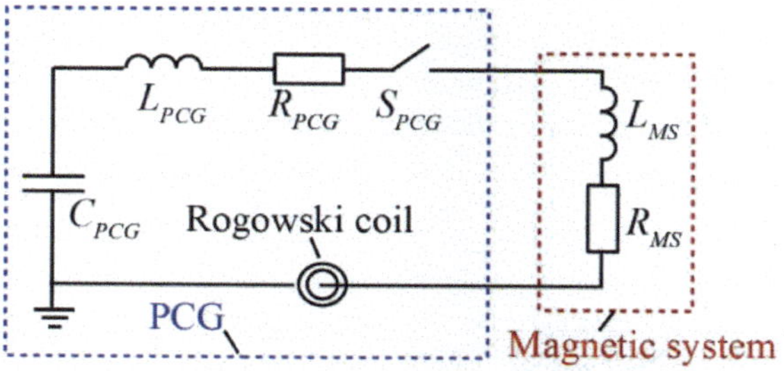

Fig. 1.1 The electrical circuit of the PCG with connected magnetic system

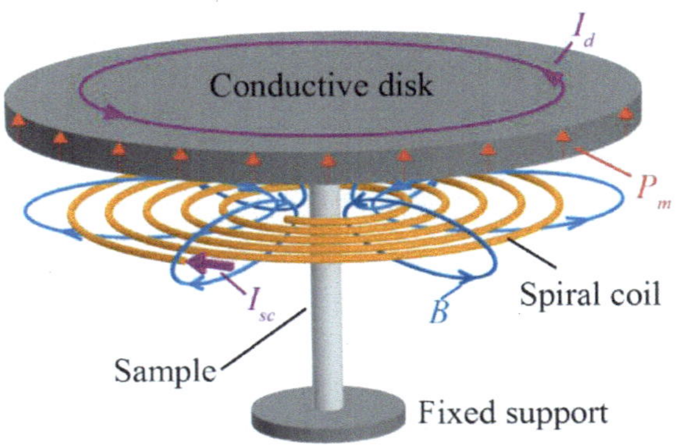

Fig. 1.2 Scheme of the uniaxial tension using spiral coil near conductive disk

In this work, we used capacitor bank with capacitance $C_{PCG} = 14.5$ µF, self-inductance of the PCG $L_{PCG} = 100$ nH, PCG oscillation period in the short circuit experiments was $T = 7.5$ µs. S_{PCG} is the multichannel low inductance switch [19].

To measure the pulsed current, a Rogowski coil was used with a time constant significantly exceeding the period of current oscillations.

We discharged PCG to various magnetic systems, which created pulsed magnetic pressure for uniaxial tension of materials (see Sects. 1.2.2, 1.2.3).

1.2.2 Uniaxial Tension Method Using Spiral Coil Near Conductive Disk

PCG discharged to the spiral coil, which is placed near conductive disk, see Fig. 1.2. Current of the coil I_{sc} generates magnetic field between the coil and the conductive disk. According to the Faraday's law induced currents I_d are created in the disk. Interaction between induced currents in the disk and magnetic field leads to the Lorentz force, integration along the thickness of the disk gives and magnetic pressure P_m makes the disk to move. Moving disk tensile the sample, which is connected to the disk at one end. Another end of the sample is fixed.

In our experiments the spiral coil consists of 18 turns, cross-section of one turn was 4 mm^2. Samples were made of copper and aluminum wires (diameter 1–2 mm) and flat bone type samples (thickness 0.5–1 mm).

1.2.3 Uniaxial Tension Method Using Flat Conductors

In this loading schemes, flat closely spaced copper conductors were placed in the sample gaps, see Fig. 1.3. PCG was discharged to the flat conductors, and current in

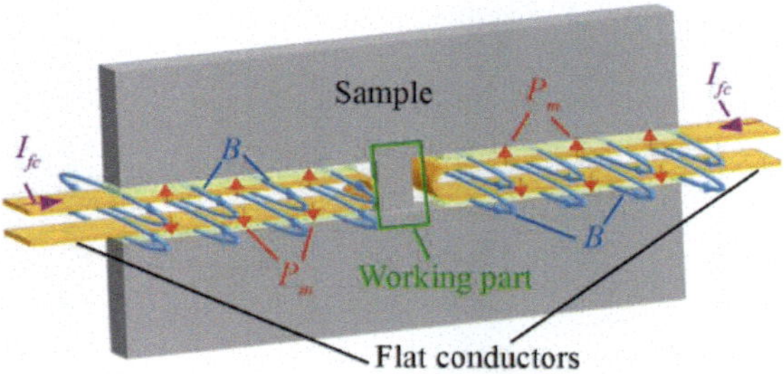

Fig. 1.3 Scheme of the uniaxial tension using flat conductors

the flat conductors I_{fc} generated pulsed magnetic field B between them. Interaction between currents of the flat conductors and magnetic field leads to the Lorentz force, integration of which along thickness of the flat conductor gives magnetic pressure, which is transferred to the sample and tensile the working part of the sample.

In our experiments thickness of the flat copper conductor was 0.3 mm, wide 4 mm. The experiment samples were made of TiNi alloy by EDM with the dimensions: length 30 mm, height 14.5 mm, thickness 2 mm. Height, width and thickness of the working part were 2.5, 1, 1 mm.

Flat conductors can be made in the quasi-coaxial form to significant reduction of induced currents in the sample, as was done for the fracture test (mode I) of the copper samples [18].

Modification of the Tension Method Using Flat Conductors. For simplifying the production of the samples, we can place the simple sample between two elastic plates, see Fig. 1.5. In this scheme mechanical stress lower than the yield stress of the plates, that simplify the analysis of the deformation process of the sample.

For these experiments, two reinforcing steel plates with dimensions of 35 mm wide, 6 mm high and each plate 1.8 mm thick were used. The thickness of the bone type specimen was 0.5 mm of oxygen-free high thermal conductivity (OFHC) copper.

1.2.4 Finite Element Modelling

Electro-Magnetic Part. *In the electric part of the conductive currents, we considered derivative form of the current conservation law, Ohm's law and dependence of electric field E on electric potential V:*

$$\begin{cases} \nabla \cdot \overline{J}_{ec} = 0, \\ \overline{J}_{ec} = \sigma \overline{E}_{ec} + \overline{J}_{e,ec}, \\ \overline{E}_{ec} = -\nabla V; \end{cases} \tag{1.1}$$

where σ is electric conductivity, J, J_e are full and external currents, index *Ec* means parameters of the electric currents part.

These laws reduce to one equation according to the electric potential V, which we solved using FEM in the Comsol Multiphysics software, module AC/DC - electric current. Knowing the electric potential, we can obtain all parameters of eclectic field.

In the magnetic part of the task, we considered derivative form of Ampere's, Faraday's and Gauss laws:

$$\begin{cases} \nabla \times \overline{H} = \overline{J}_{mf}, \\ \overline{B} = \nabla \times \overline{A}, \\ \overline{J}_{mf} = \sigma \overline{E}_{mf} + \overline{J}_{e,mf}, \\ \overline{E}_{mf} = -\dfrac{\partial \overline{A}}{\partial t}; \end{cases} \tag{1.2}$$

where H is magnetic field strength, B is magnetic flux density, index mf means parameters of the magnetic field.

These laws reduce to the one equation according to the magnetic vector potential A, which we solved using FEM in the Comsol Multiphysics software, module AC/DC—magnetic fields. Knowing the vector magnetic potential, we can obtain all parameters of a magnetic field.

Coupling between electric and magnetic fields done throw external current densities: conductive current from electric part transferred to magnetic part and induced currents from magnetic part transferred to the electric part:

$$\begin{cases} \overline{J}_{e,mf} = \sigma \overline{E}_{ec} = -\sigma \nabla V, \\ \overline{J}_{e,ec} \overline{E}_{mf} = -\sigma \dfrac{\partial \overline{A}}{\partial t}. \end{cases} \tag{1.3}$$

Thermal Part. Knowing the current densities from electro-magnetic part we calculated Joule's heating power Q and temperature T distribution in the samples, according to the law of the energy conservation in derivative form, in Comsol software, module heat transfers in solids:

$$Q = |\overline{J}|^2 / \sigma \tag{1.4}$$

$$\rho C \frac{\partial T}{\partial t} - \nabla \cdot k \nabla T = Q \tag{1.5}$$

where ρ is mass density, C is heat capacitance, k is thermal conductivity.

Mechanical Part. Knowing the current densities and magnetic flux densities from electro-magnetic part we calculated Lorentz forces f_v and magnetic pressure by integration of this forces. Then we applied this pulsed pressure as boundary condition for FEM of the derivative form of the second Newton's law in the ANSYS AutoDyn software:

$$\overline{f}_v = \overline{j} \times \overline{B} \tag{1.6}$$

$$\rho \frac{\partial^2 u}{\partial t^2} - \nabla \cdot \ddot{\sigma}_m = f_v \tag{1.7}$$

where u is displacements, σ_m is mechanical stress tensor.

Elastic part of the material described by Young law, plasticity was described using Johnson–Cook model for copper OFHC samples and multi-linear hardening law for TiNi samples.

1.3 Results

1.3.1 Calculation of the Magnetic Field

Flat Conductors. Using FEM, we solved the magnetic field of the simple flat conductors and quasi-coaxial conductors, see Fig. 1.6.

As you can see that using of quasi-coaxial conductors allows to reduce significantly magnetic field and induced currents in the sample. This effect should be considered as positive, because in some cases high level of the induced current in the samples can lead to the additional Lorentz forces and heating of the samples, which can affect on deformation process.

We calculated magnetic pressure by integration of the Lorentz force along thickness of the flat conductors. In stationary mode this calculation gives the same results like analytical solution [20].

Spiral Coil Near Conductive Disk. Using FEM, we solved the magnetic field of the spiral solenoid near conductive disk, see Fig. 1.7.

You can see than the highest level of the magnetic field placed between spiral coil and conductive disk. We calculated magnetic pressure by integration of the Lorentz force inside the disk.

1.3.2 Mode I Fraction Test Using Flat Conductors

We performed series of the experiments on mode I fracture test using quasi-coaxial flat conductors, according to the Fig. 1.4. The currents of flat conductors you can see on the Fig. 1.8.

After experiments we measured the residual strain of the samples in the top of the gap, see Fig. 1.9, and compare it with FEM in ANSYS AutoDyn using Johnson–Cook plasticity model for OFHC copper.

You can see good correlation between experimental and calculation data up to magnetic pressure 400 MPa. Higher level of the magnetic pressure (> 400 MPa)

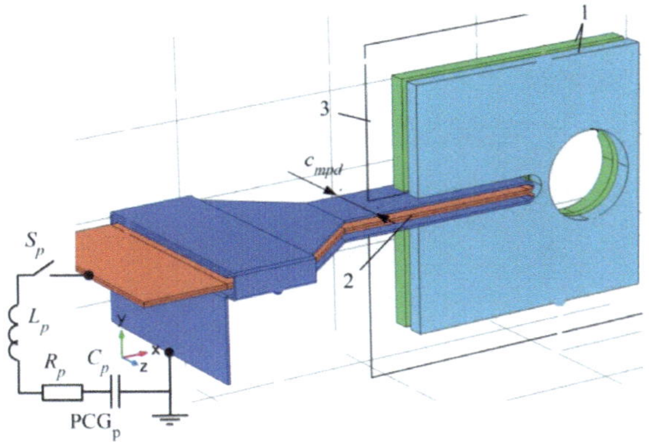

Fig. 1.4 Scheme of the fracture test (mode I) using flat conductors: 1—two samples; 2—quasi-coaxial flat conductors; 3—insulation

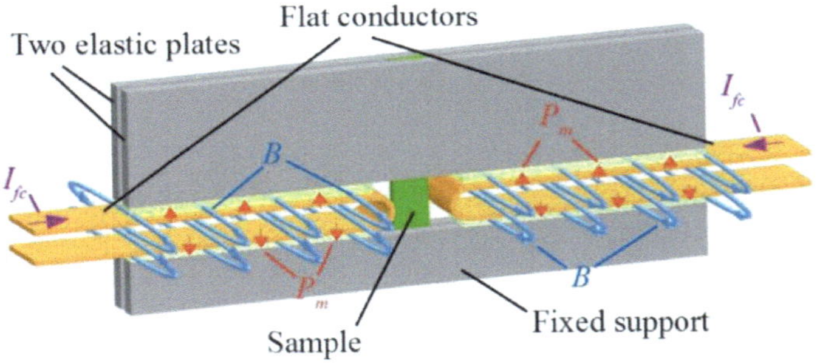

Fig. 1.5 Modification of the scheme of the uniaxial tension using flat conductors

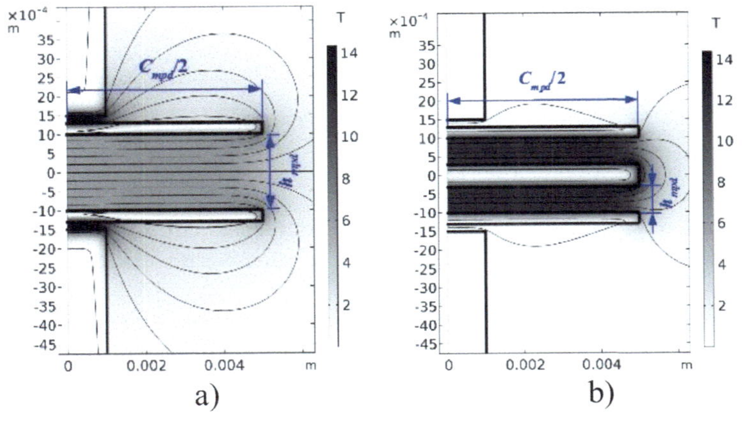

Fig. 1.6 Distribution of the magnetic field of the flat conductors: **a** simple flat conductors; **b** quasi-coaxial flat conductorUniaxial Tension Method Using Fls

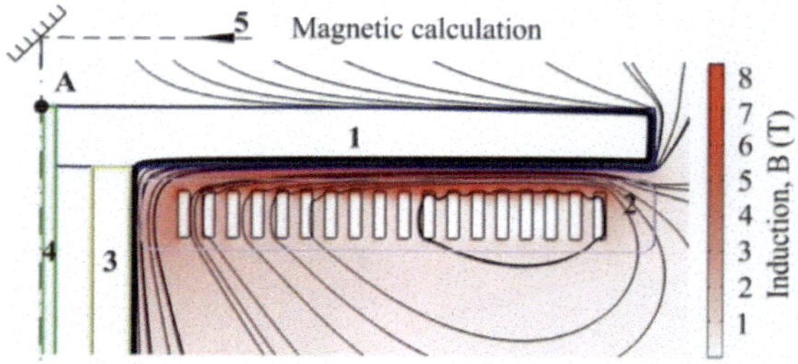

Fig. 1.7 Distribution of the magnetic field of the spiral coil

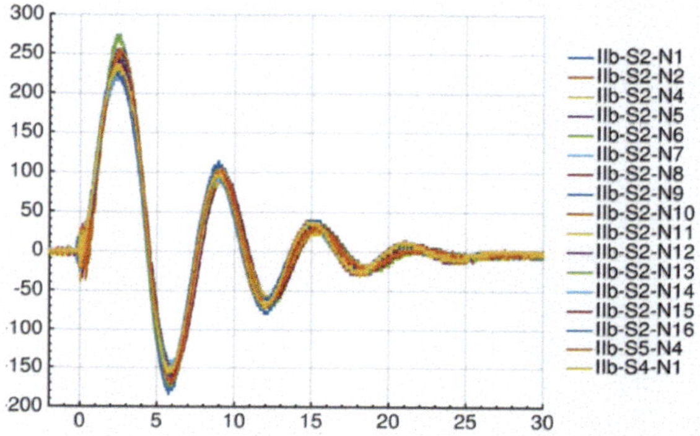

Fig. 1.8 Current of the quasi-coaxial flat conductors in the mode I fraction test of the copper OFHC samples

leads to the fracture of the samples. Strain rate according to the calculation was in the range 2000–7000 1/s.

1.3.3 Uniaxial Tension Method Using Spiral Coil Near Conductive Disk

We performed series of the experiments on uniaxial tension of wire and bone type samples, see Fig. 1.10, using spiral coil near conductive disk, according to the Sect. 1.2.2.

In these experiments we used high speed camera for measuring the deformations. The strain rate was up to 2500 1/s. More about these experiments you can read in [21].

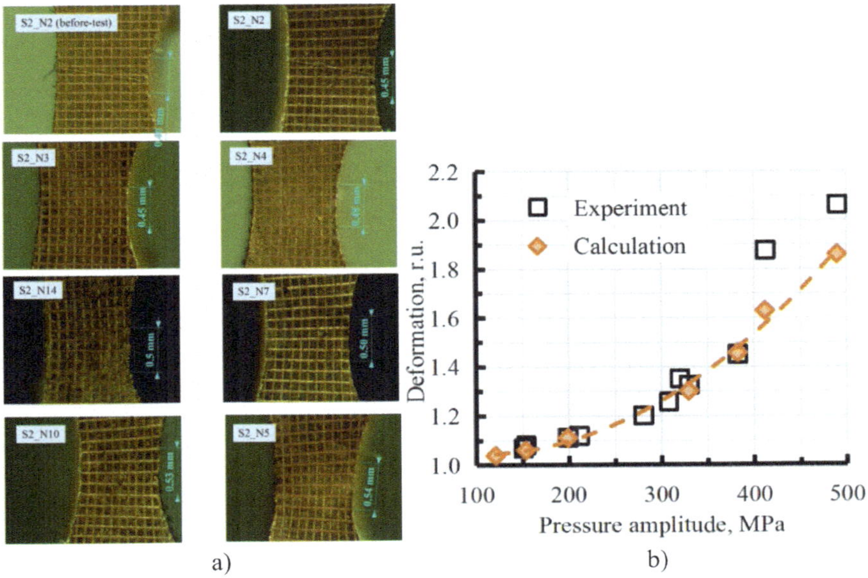

Fig. 1.9 **a** Residual strains in the top of the sample gap; **b** dependence of residual strain in the top of the sample gap on magnetic pressure amplitude

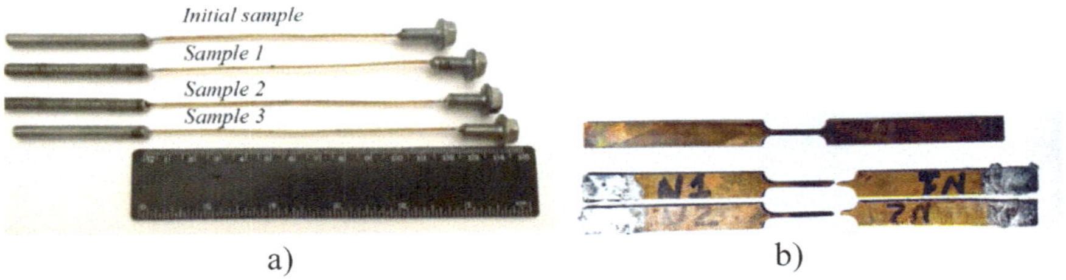

Fig. 1.10 Samples of the uniaxial tension using spiral coil near conductive disk: **a** copper wire samples; **b** copper bone type copper samples

1.3.4 Uniaxial Tension Method Using Flat Conductors

We performed series of the experiments on uniaxial tension of TiNi samples, see. Figure 1.11, using a couple of flat conductors, according to the Sect. 1.2.3.

In these experiments we used laser and photodetector for measuring the deformations. We also calculated mechanical deformation of the samples in ANSYS AutoDyn software using multi-linear hardening model and compare the results with residual strain of the samples, see Fig. 1.12.

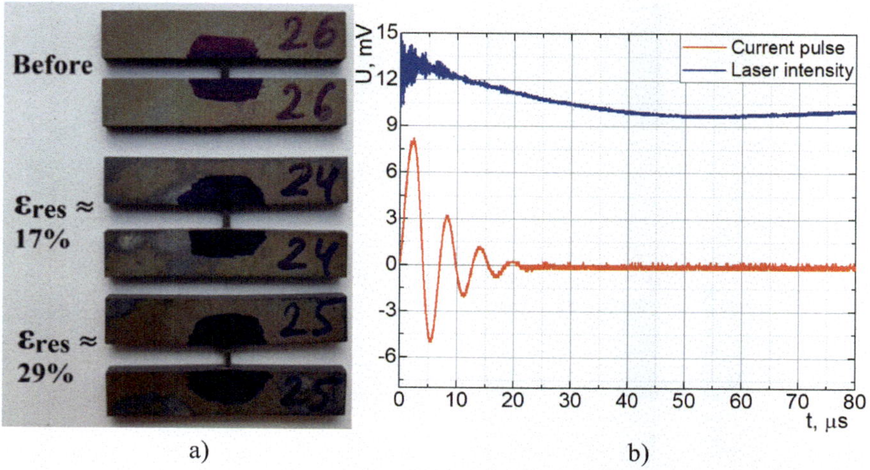

Fig. 1.11 Uniaxial tension method using flat conductors: **a** Samples of the TiNi alloy; **b** Dependences of the experiment current and photodetector signal

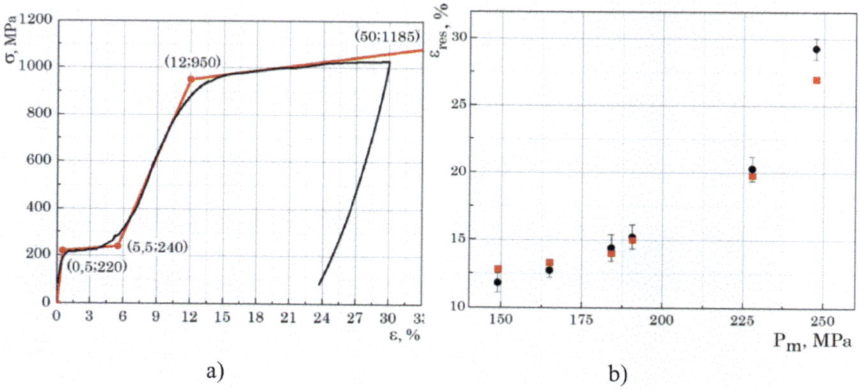

Fig. 1.12 **a** Multi-linear hardening model of the TiNi alloy; **b** Experiment (black points) and calculated (red points) of the samples' residual strains

The strain rate was in the range 2000–6000 1/s. More about these experiments you can read in [22].

Calculation of the samples heating. We calculated the heating of the samples in our experiments, see Fig. 1.13.

You can see insignificant heating of the working part of the samples. It does not affect on the deformation process of the samples.

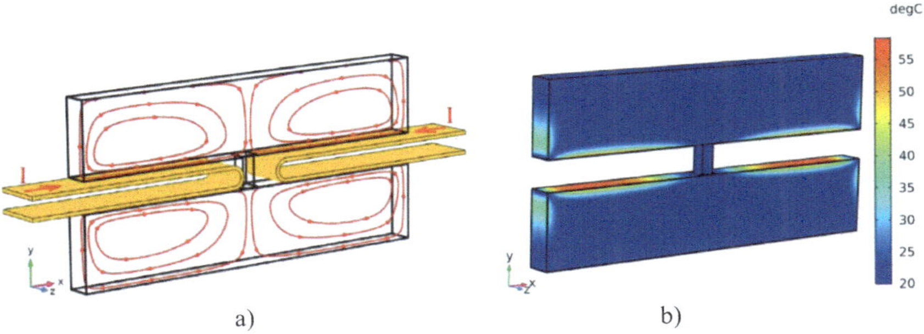

Fig. 1.13 Result of the FEM modelling of the uniaxial tension using flat conductors: **a** Stream lines of the current density; **b** distribution of the temperature

1.3.5 *Modification of the Tension Method Using Flat Conductors*

We performed the modification of the tension method using flat conductors, according to the schemes in Sect. 1.2.3, see Fig. 1.14.

FEM modeling in ANSYS AutoDyn of the mechanical deformation shows that mechanical stresses in reinforcing steel plates lower than yield stress. And low mass

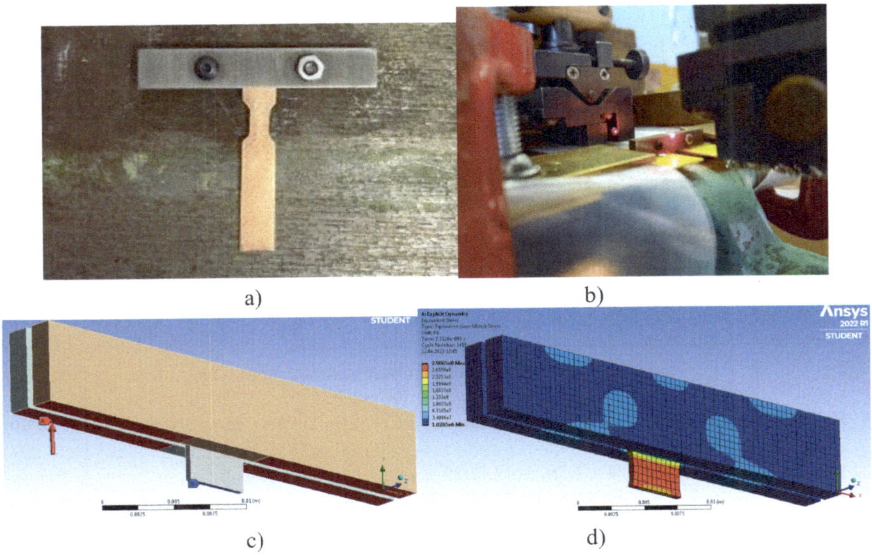

Fig. 1.14 Modification of the tension method using flat conductors: **a** the OFHC copper sample with the loading plates; **b** experimental setup; **c** FEM model (red boundaries—external pressure; blue boundary—fixed); **d** distribution of the Von Mises stress

of the plates $m = 6$ g gives good sensitivity for the FEM verification of plasticity models of the samples: allows to register a 10% change in the parameters of the Johnson–Cook high strain rate plasticity model, using copper OFHC samples as an example.

1.4 Conclusion

The use of a pulsed magnetic field makes it possible to form impact loads for testing materials by various methods. In this paper we considered the possibility for uniaxial high strain rate tension of the materials using spiral coil and flat conductors.

Modification of uniaxial tension using flat conductors, which allows to simplify analysis of the experimental results was described.

Registration of the deformation process using high speed camera or laser with photodetector gives data on the testing samples strain. Using FEM method, we can verify plasticity models. But it would be more convenient if the stress of the samples was measured during the experiments, that is the next step of presented study of high strain rate deformation process.

Acknowledgements This research was done by Peter the Great St. Petersburg Polytechnic University and supported under the strategic academic leadership program 'Priority 2030' of the Russian Federation (Agreement 075-15-2021-1333 dated 30.09.2021);

Primary uniaxial tension scheme using flat conductors and experiments on TiNi alloy specimens was funded by RFBR, project number 19-32-60035;

Modification of the the uniaxial tension method using flat conductors done by Grant of the President of the Russian Federation for state support of young Russian scientists—candidates of sciences MK-2816.2022.4.

References

1. Johnson GR, Cook WH (1985) Fracture characteristics of three metals subjected to various strains, strain rates, temperatures and pressures. Eng Fract Mech 21:31–48. https://doi.org/10.1016/0013-7944(85)90052-9
2. Armstrong RW, Zerilli FJ (1987) Dislocation-mechanics-based constitutive relations for material dynamics calculations. J Appl Phys 61:1816–1825
3. Follansbee PS, Kocks UF (1988) A constitutive description of the deformation of copper based on the use of the mechanical threshold stress as an internal state variable. Acta Metall 36:81–93. https://doi.org/10.1016/0001-6160(88)90030-2
4. Preston DL, Tonks DL, Wallace DC (2003) Model of plastic deformation for extreme loading conditions. J Appl Phys 93:211–220. https://doi.org/10.1063/1.1524706
5. Taylor G (1948) The use of flat-ended projectiles for determining dynamic yield stress. Proc R Soc London A 194:289–299
6. Kolsky H (1949) An investigation of the mechanical properties of materials at very high rates of loading. Proc Phys Soc Sect B 62:676–700. https://doi.org/10.1088/0370-1301/62/11/302

7. Dharan CKH, Hauser FE (1970) Determination of stress-strain characteristics at very high strain rates. Exp Mech 10:370–376. https://doi.org/10.1007/bf02320419
8. Bragov AM, Balandin VV, Konstantinov AY, Lomunov AK, Vorobtsov IV, Kuznetsov AV, Savenkov GG (2017) High-rate deformation and spall fraction of some metals. Procedia Eng 197:260–269
9. Zhang H, Ravi-Chandar K (2006) On the dynamics of necking and fragmentation—I. Real-time and post-mortem observations in Al 6061-O. Int J Fract 142:183–217. https://doi.org/10.1007/s10704-006-9024-7
10. Zhang H, Ravi-Chandar K (2008) On the dynamics of necking and fragmentation—II. Effect of material properties, geometrical constraints and absolute size. Int J Fract 150:3–36. https://doi.org/10.1007/s10704-008-9233-3
11. Morozov V, Kats V, Savenkov G, Lukin A (2018) Mechanisms of fracture of ring samples made of FCC metals on loading with magnetic-pulse method. AIP Conf Proc 1959. https://doi.org/10.1063/1.5034753
12. Gruzdkov A, Krivosheev S, Petrov Y, Razov A, Utkin A (2008) Martensitic inelasticity of TiNi-shape memory alloy under pulsed loading. Mater Sci Eng A 481–482:105–108. https://doi.org/10.1016/j.msea.2007.03.113
13. Nie H, Suo T, Wu B, Li Y, Zhao H (2018) A versatile split Hopkinson pressure bar using electromagnetic loading. Int J Impact Eng 116:94–104. https://doi.org/10.1016/j.ijimpeng.2018.02.002
14. Guo Y, Du B, Liu H, Ding Z, Zhao Z, Tang Z, Suo T, Li Y (2020) Electromagnetic Hopkinson bar: a powerful scientific instrument to study mechanical behavior of materials at high strain rates. Rev Sci Instrum 91:0–10. https://doi.org/10.1063/5.0006084
15. Krivosheev SI (2005) Pulsed magnetic technique of material testing under impulsive loading. Tech Phys 50(3):334–340
16. Avriel E, Lovinger Z, Nemirovsky R, Rittel D (2018) Investigating the strength of materials at very high strain rates using electro-magnetically driven expanding cylinders. Mech Mater 117:165–180. https://doi.org/10.1016/j.mechmat.2017.11.009
17. Deng H, Yang S, Li G, Zhang X, Cui J (2020) Novel method for testing the high strain rate tensile behavior of aluminum alloys. J Mater Process Technol 280. https://doi.org/10.1016/j.jmatprotec.2020.116601
18. Magazinov SG, Krivosheev SI, Adamyan YE, Alekseev DI, Titkov VV, Chernenkaya LV (2018) Adaptation of the magnetic pulse method for conductive materials testing. Mater Phys Mech 40:117–123. https://doi.org/10.18720/MPM.4012018_14
19. Adam'yan YuE, Vasilevskij VM, Krivosheev SI, Shutov VL (1995) Low-inductive multichannel gas current commutator. Prib Tekh Eksp 2:122–125
20. Ostropiko ES, Krivosheev SI, Magazinov SG (2021) Analytical evaluation of magnetic pulse deformation of tini alloy. Lett Mater 11:55–60. https://doi.org/10.22226/2410-3535-2021-1-55-60
21. Krivosheev SI, Magazinov SG, Adamyan YE, Alekseev DI, Manzuk MV (2020) Uniaxial high strain rate tension with the use of magnetic pulse method. Proc. 2020 IEEE conference of Russian young researchers electrical and electronic engineering EIConRus, pp 1082–1086. https://doi.org/10.1109/EIConRus49466.2020.9039159
22. Ostropiko E, Magazinov S, Krivosheev S (2022) Uniaxial magnetic pulse tension of TiNi alloy with experimental strain rate evaluation. Exp Mech 62:1027–1036. https://doi.org/10.1007/s11340-022-00864-4

Chapter 2
About the Shift of the Deformation Creep Curves of Carbon Fiber Reinforced Plastics After Aging for Different Programs

Alexander R. Arutyunyan

Abstract At recent decades, polymer composite materials become more widespread and developed in critical areas of engineering practice. They are used, in particular, in aircraft, rocket engineering, shipbuilding, railway transport and construction. Advantages of these materials are high strength and stiffness, lightness, high fatigue strength, chemical inertness, heat and electrical conductivity, and others. Because these materials are used in critical areas of engineering practice, so this makes their long-term aging and creep characteristics of paramount importance. At the same time, the physical and chemical characteristics of these materials essentially changed after long-term operation, significantly due to the aging process. Thus, investigations of aging of these materials are much needed. Experimental studies on alternation of natural and thermal aging and deformation aging were conducted to study the evolution of creep characteristics of carbon fiber reinforced plastics (CFRP). The experimental creep curves after different aging programs were received. In total experiments were carried out during 3 years. It is shown, that the aging process has non-monotonic character, which should be considered in engineering practice at design of structural elements made of polymer composite materials.

2.1 Introduction

Polymers and polymer-based composite materials are intensive implemented in almost all areas of industry and construction, in particular, in aircraft, rocket and automobile construction, shipbuilding, railway transport, agricultural machinery production, tidal power, as well as in the manufacture of sport equipment due to their high strength to weight ratio, superior fatigue strength, and design flexibility. These applications frequently have a very high-performance demand, which makes their long-term characteristics of paramount importance. At the same time, they were subject

A. R. Arutyunyan (✉)
St. Petersburg State University, Universitetsky Pr., 28, St. Petersburg, Petrodvorets 198504, Russia
e-mail: a.arutyunyan@spbu.ru

H. Altenbach and V. Kuzkin (eds.), *Advanced Problems in Mechanics*,
https://doi.org/10.1007/978-3-031-80103-7_2

to degradation due to environmental factors, including light, temperature, stress, and others. During the aging, the degradation of polymer and composite materials leads to significantly changes in the physical and mechanical properties [1–5].

In experiments polymer resin, polymer and based on them composites materials are usually tested during long duration in a liquid medium at certain temperature conditions. This method to evaluate polymer and composites materials service conditions is termed as hygrothermal aging. The hygrothermal aging at high temperature accelerates the aging process. The natural seawater, distilled water, tap water, demineralized water, alkaline and acid environment, salt solution or any other chemical based applications can be used as aging agents. The influence of aging on the mechanical properties of the fiber reinforced plastics at various environmental factors and loading conditions are investigated in works [6–8].

The paper [6] reports the recently found critical drawback of glass fibers reinforced plastics (GFRP) as compared with carbon fibers reinforced plastics (CFRP) composites. GFRP and CFRP composites were placed into a general water bath at 80 °C until saturation for accelerated hygrothermal ageing. After ageing, the impact properties of CFRP composites improved. For GFRP composites, however, their performance under low-velocity impact degraded significantly. After removing the moisture absorbed during ageing, the strength of CFRP laminates increased to 95.75% of its original value. In contrast, GFRP composites only retained 74.65% of the strength after hygrothermal treatment.

In [7] the mechanical properties degradation of CFRP under different service conditions of marine environment were studied. In the vertical fiber direction, the marine environment will significantly reduce the tensile strength and elastic modulus of CFRP, and further decrease with the increase of temperature; while in the fiber direction, it has little effect on the initial mechanical properties of CFRP, and the maximum decrease of elastic modulus is only 4.45%. There is little difference between the pure water immersion and salt immersion environments on the initial performance of CFRP. Pure water and salt immersion environment have the same effect on the creep properties of CFRP. The microstructure analysis shows that water molecules and NaCl solution improved the mobility of epoxy molecular fragments, thereby accelerating the creep failure of CFRP. The microscopic energy analysis shows that the transformation process of CFRP strain energy before failure and the deconvolution of epoxy molecular chain after failure have great influence on the creep behavior. The decrease of interface properties caused by the increase of interfacial energy of CFRP in marine environment is also an important cause of creep failure.

The study [8] examined the prediction of statistical life time for developed carbon fiber reinforced thermoplastic strands (CF/TP strands) at creep and fatigue tension loadings under the wet condition of water absorption with comparison to similar prediction under a dry condition. Finally, the influences of water absorption on the statistical long term creep and fatigue strengths of CF/TP strands are cleared. In particular, the degradation of relaxation modulus of matrix thermoplastic epoxy (TP-EP) resin with increasing of elapsed time is accelerated with the water absorption, and the degradation of creep strength of CF/TP strand with increasing of elapsed

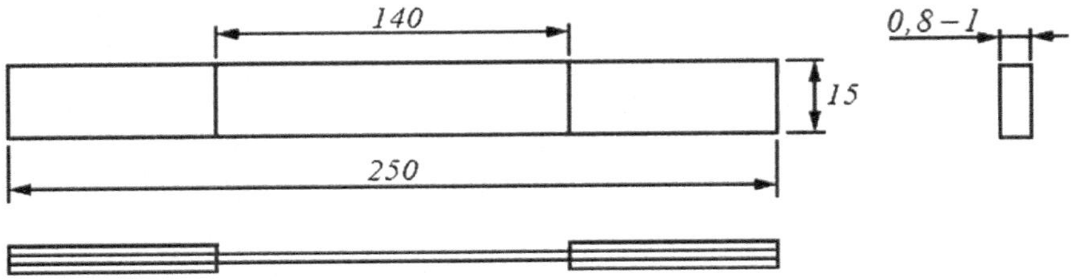

Fig. 2.1 Geometry and dimensions of specimens and tabs (in millimeters)

time is also accelerated with the water absorption in the similar manner of matrix resin. On the other hand, the fatigue strength of CF/TP strand decreases scarcely with water absorption although this fatigue strength depends remarkably on the number of cycles to failure.

In this paper experimental study on alternation of natural and thermal aging and deformation aging were conducted to study the evolution of creep characteristics of carbon fiber reinforced plastics composites.

2.2 Materials and Methods

Specimens made of carbon fiber reinforced plastic (CFRP) composite of the T26/22502/I131636 mark with a length of 250 mm, a working length of 140 mm, a width of 15 mm and a thickness of 0.8–1 mm we used in experiments. The geometry and dimensions of specimens and tabs are shown on Fig. 2.1. Dimensions are given in millimeters.

The experiments on creep were performed on a tearing test machine TINIUS OLSEN H10K-T at room temperature. Aging at evaluated and low temperatures were conducted correspondingly in muffle furnace Nabertherm N40E and in the standard freezer.

2.3 The Programs of Climatic, Thermal and Deformation Aging of CFRP Composite Specimens

To study the evolution of creep characteristics of carbon fiber reinforced plastic composites the experiments on alternation of natural and thermal aging, cyclic loadings and deformation aging were conducted on CFRP specimens of the T26/22502/I131636 mark. In total experiments were carried out during 3 years.

The aging of CFRP composite specimens was carried out according to the following programs.

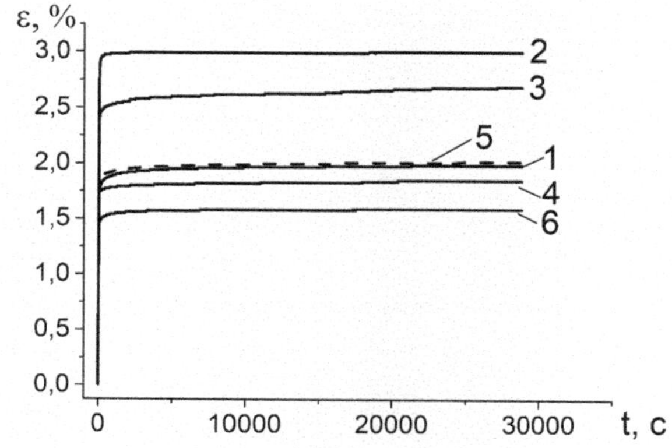

Fig. 2.2 Experimental creep curves for specimens of CFRP composite of the T26/22502/I131636 mark with aging according program 1 (curves 1–4), program 2 (curve 5) and program 3 (curve 6) under constant stress equal to 240 MPa (0,42 of ultimate tensile strength)

Program 1. Artificial aging in ice at –16 °C for 2 days, then exposure in distilled water for 3 weeks; aging in the laboratory conditions during 5 months; creep experiments at $\sigma = 240$ MPa (0,42 of the tensile strength) for 8 h (curve 1 on Fig. 2.2); exposure in distilled water for 3 weeks, then artificial aging in ice at –16 °C for 3 weeks; aging in the laboratory conditions during 4 months; creep experiments at $\sigma = 240$ MPa for 8 h (curve 2 on Fig. 2.2); aging in the laboratory conditions during 7 months; cyclic loadings with a constant amplitude with a sinusoidal stress change $\sigma = 276{,}35$ MPa (approximately 0,5 of the tensile strength), a cycle asymmetry coefficient R = 0 and loading frequency of 20 Hz up to 15 000 000 cycles; aging in the laboratory conditions during 4 months; creep experiments at $\sigma = 240$ MPa for 8 h (curve 3 on Fig. 2.2); aging in the laboratory conditions during 7 months; exposure in distilled water for 5 days, then artificial aging in ice at –16 °C for 2 month, exposure in distilled water for 7 days; aging in the laboratory conditions during 4 months; creep experiments at $\sigma = 240$ MPa for 8 h (curve 4 on Fig. 2.2).

Program 2. Artificial aging in ice at –16 °C for 2 days, then exposure in distilled water for 3 weeks; aging in the laboratory conditions during 7 months; exposure in distilled water for 3 weeks, then artificial aging in ice at –16 °C for 3 weeks; aging in the laboratory conditions during 11 months; exposure in distilled water for 5 days, then artificial aging in ice at –16 °C for 2 month, exposure in distilled water for 7 days; aging in the laboratory conditions during 2 months; creep experiments at $\sigma = 240$ MPa for 8 h (curve 5 on Fig. 2.2).

Specimens were tested only on alternation of natural and thermal aging at negative temperatures. Therefore, program 2 is similar to program 1 only without creep and fatigue experiments in the intermediate stages.

Program 3. Artificial aging at 120 °C for 2 h, then exposure in distilled water for 3 weeks; aging in the laboratory conditions during 7 months; exposure in distilled water for 3 weeks, then artificial aging at 120 °C for 2 h; aging in the laboratory conditions during 23 months; creep experiments at $\sigma = 240$ MPa for 8 h (curve 6 on Fig. 2.2).

Specimens were tested only on alternation of natural and thermal aging at high temperatures. Therefore, program 3 is similar to program 2 only the exposure of specimens was carried out at elevated temperature.

2.4 Experimental Results on the Influence of Climatic, Thermal and Deformation Aging on Creep Curves of CFRP Composite Specimens

On the Fig. 2.2, the experimental creep curves for specimens of CFRP composite of the T26/22502/I131636 mark with aging according program 1 (curves 1–4), program 2 (curve 5) and program 3 (curve 6) under constant stress equal to 240 MPa (0,42 of ultimate tensile strength) are shown.

According program 1 at the first stages of deformation aging a hardening effect during creep by 1.5 times by deformation level is occurred (curve 2 on Fig. 2.2). Then, after additional aging the softening is observed (curve 3 on Fig. 2.2) and deformation is equal to only 1.35 from initial deformation without aging. After additional aging, the softening effect continues to be observed (curve 4 on Fig. 2.2) and the deformation falls below the deformation without aging (it is equal to only 0.93 from deformation without aging). Therefore, the process of aging has non-monotonic character, which should be considered in engineering practice at design of structural elements made of polymer composite materials.

It is can be seen, that aging according program 2 has no effect on creep characteristics and creep deformation in this case is equal to initial deformation without aging (curve 5 on Fig. 2.2).

Aging program 3 significantly effect on creep characteristics (curve 6 on Fig. 2.2) and creep deformation in this case is equal to only 0.8 from initial deformation without aging.

2.5 Conclusions

Experiments to study the evolution of creep characteristics of carbon fiber reinforced plastic (CFRP) of the T26/22502/I131636 mark were conducted. In total experiments were carried out during 3 years. At program 1 specimens were tested on alternation of creep, natural aging, thermal aging at negative temperatures and cyclic loadings. At the first stages of deformation aging a hardening effect during creep by 1.5 times is occurred. Then, after additional aging the softening is observed. Therefore, the process of aging has non-monotonic character, which should be considered in engineering practice at design of structural elements made of polymer composite materials.

At program 2 specimens were tested only on alternation of natural and thermal aging at negative temperatures. In total experiments were carried out during 3 years. Then specimens were tested on creep. It is shown, that such aging program has no effect on creep characteristics and creep deformation in this case is equal to initial deformation without aging.

At program 3 specimens were tested only on alternation of natural and thermal aging at high temperatures. In total experiments were carried out during 3 years. Then specimens were tested on creep. It is shown, that such aging program significantly effect on creep characteristics and creep deformation in this case is equal to only 0.8 from initial deformation without aging.

Acknowledgements Financial support of the Russian Science Foundation is gratefully acknowledged (Grant N 23-29-00383).

References

1. Struik LCE (1978) Physical aging in amorphous polymers and other materials. Elsevier Sci Publ Comp, Amsterdam, Oxford, New York
2. Nadai A (1963) Theory of flow and fracture of solids. McGraw-Hill, New York
3. Hutchinson JM (1995) Physical aging of polymers. Prog Polym Sci 20(4):703–760
4. Arutyunyan RA (2004) The problem of strain aging and long-term fracture in mechanics of materials. Publishing House of the St, Petersburg State University, St. Petersburg (in Russian)
5. Davies P, Rajapakse YDS (2018) Durability of composites in a marine environment 2. Solid mechanics and its applications 245. Springer International Publishing
6. Zhong Y, Cheng M, Zhang X, Hu H, Cao D, Li SH (2019) Hygrothermal durability of glass and carbon fiber reinforced composites a comparative study. Compos Struct 211:134–143
7. Li X, Zhang X, Chen J, Huang L, Lv Y (2022) Effect of marine environment on the mechanical properties degradation and long-term creep failure of CFRP. Mater Today Commun 31:103834
8. Nakada M, Miyano Y, Kageta S, Nishida H, Hayashi Y, Uzawa K (2022) Prediction of statistical life time for unidirectional CFRTP under water absorption. J Reinf Plast Compos 41(7–8):257–266

Chapter 3
About the Possibility of Experimental Creep Curves Using to Determine the Metallic Materials Damage

Alexander R. Arutyunyan and Regina R. Saitova

Abstract Under the action of relatively low stresses and high temperatures, metallic materials become brittle and fracture with a small value of residual deformations. This problem is known as a problem of thermal brittleness of metals. To solve this problem in the mechanics of materials the damage conception was introduced. To describe the brittle region of the experimental long-term strength curve, the system of simple kinetic equations for the damage parameter and creep deformation was proposed, and the long-term strength criterion was formulated. In this work we proposed to determine the damage parameter changes according to the experimental high-temperature creep curves. Only one kinetic equation for creep rate for compressible medium recorded using the damage parameter in the form of relative density changes is formulated. From this equation the damage parameter is determined depending on the creep rate and the creep deformation. To describe the experimental creep curves various empirical dependences in the form of power, exponential and mixed functions are used. Theoretical damage curves are plotted. The corresponding theoretical long-term strength curves are constructed. Comparison of theoretical creep curves with experimental results for 2.25Cr-1Mo steel at 500°C, 12% Cr-X20 steel at 549.85°C and P91 steel at 600°C is given. A good agreement between the corresponding theoretical and experimental results is observed.

3.1 Introduction

The problem of thermal brittleness is considered, when under the action of relatively low stresses and high temperatures, metallic materials become brittle and fracture with a small value of residual deformation. The damage conception that was introduced in the mechanics of materials to describe long-term strength under conditions of high-temperature creep, has been developed in the fundamental works of Kachanov [1] and Rabotnov [2]. In their works, to describe the brittle region of the

A. R. Arutyunyan (✉) · R. R. Saitova
St. Petersburg State University, Universitetsky Pr., 28, St. Petersburg, Peterhof 198504, Russia
e-mail: a.arutyunyan@spbu.ru

H. Altenbach and V. Kuzkin (eds.), *Advanced Problems in Mechanics*,
https://doi.org/10.1007/978-3-031-80103-7_3

experimental long-term strength curve, the simple kinetic equation for the damage parameter was proposed, and the long-term strength criterion was formulated. The question of the relationship between creep deformation and damage in these works was not discussed.

The next stage on the creep and damage problem solution relates to the work of Rabotnov [3], in which a system of two interrelated equations for creep deformation and damage parameter was proposed.

Nowadays in our early works we proposed a modified version of the system of kinetic equations of Kachanov-Rabotnov [4, 5]. The mass conservation law was taken into account and the damage parameter was specified in the form of a relative changes of the material density.

In the scientific literature on this problem, the following possible variants of the relationship between creep and damage are given. The processes of creep and damage develop in parallel and in the first approximation do not relate to each other. Damage is the result of deformation, which creates fracture sources, leads to appearance of places with a high stress concentration and is a producer of point defects that are necessary for the development of slow fracture. Creep is a result of the micro-fracture processes in the material volume.

Low deformation rates and high temperatures contribute to intergranular creep fracture. This fact let assume that damage and fracture could flow independently to plastic deformation. This is also referred by numerous cases of slow fracture with a very small value of the residual deformation. Studies of Ratcliffe and Greenwood [6], Betechtin [7] on the density changes in creep conditions have shown that the pores healing by a single and multiple application of hydrostatic pressure leads to a sharp slowing down of creep deformation. Therefore, the time to fracture significantly increases. At the same time, the creep rate practically does not change. The processes of damage by density changes are completely slowing down the fracture development, which indicates the independence of the creep rate from damage. Apparently, all three possible variants of the relationship between creep and fracture are fair.

3.2 Description of Creep Curves

The experimental creep deformation curves can be described by various empirical dependences in the form of power, exponential and mixed functions [8–10].

In work [11], the following dependence was considered

$$\varepsilon = \frac{e^{kt}}{(f\ t + c)^n} + b \tag{3.1}$$

where f, c, k, b, n are constants.

On Fig. 3.1, the experimental creep deformation curves according to papers [12–14] and the empirical dependence in the form of function (1) are shown. In the

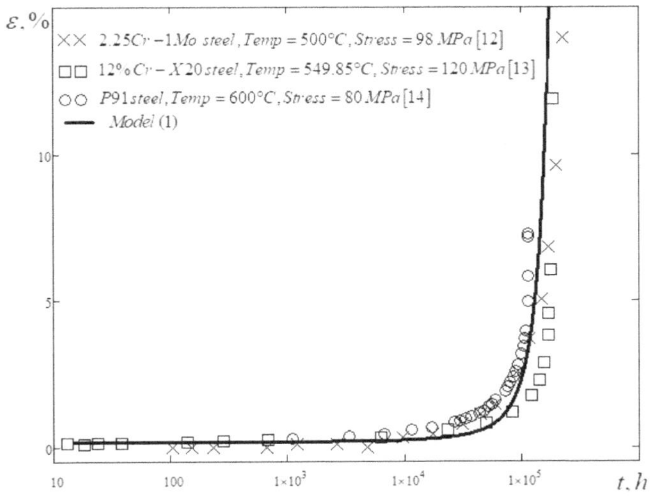

Fig. 3.1 Experimental creep deformation curves for 2.25Cr-1Mo steel at 500°C [12], 12% Cr-X20 steel at 549.85°C [13] and P91 steel at 600°C [14] and the theoretical creep deformation curve according to relation (3.1)

calculations the following coefficient values are used: $c = 10^4$, $b = 0$, $k = 2.6 \cdot 10^{-5}[h]^{-1}$, $n = 0.15, f = 0.08\,[h]^{-1}$. As can be seen from Fig. 3.1, dependence (1) describes well the experimental creep deformation curves for different steels.

3.3 Damage Parameter Changes

It is proposed to determine the damage value using experimental creep curves. The compressibility of metallic materials is taken into account and the relative density changes are considered as a damage parameter. Only one kinetic equation for creep rate is formulated. From this equation, the damage parameter is determined depending on the creep rate and the creep deformation. To describe the experimental creep curves, empirical dependence in the form of Eq. (3.1) will be used.

In our early works [4, 5] we introduced a system of interrelated kinetic equations for a compressible medium with a damage parameter defines as $\psi = \rho/\rho_0$ (ρ_0 is initial, ρ is current specimen density). Taking into account the mass conservation law, the kinetic equation for the deformation rate is given in the form

$$\frac{d\varepsilon}{dt} = \dot{\varepsilon} = B\sigma_0^m \psi^{m-\beta} e^{m\varepsilon} \tag{3.2}$$

Fig. 3.2 The theoretical damage curve according to formula (3.4)

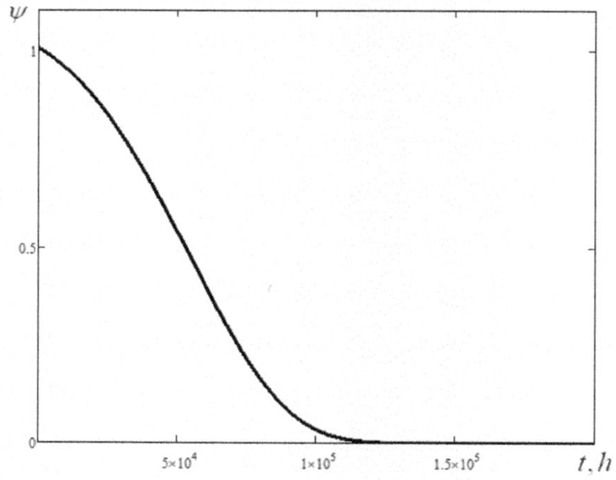

where B, m, β are constants.

From Eq. (3.2) we can receive the following equation

$$\psi = \left(\frac{\dot{\varepsilon}}{\sigma_0^m B^{m\varepsilon}}\right)^{\frac{1}{m-\beta}} \tag{3.3}$$

Using formula (3.3) we can determine the time dependence of the damage parameter according to the experimental creep deformation curves, which can be described by various empirical dependences in the form of power, exponential, and mixed functions according relation (3.1). Taking into account relation (3.3) from (1), we will have

$$\psi = \left[\frac{e^{kt}\left(k - \frac{nf}{f\ t+c}\right)}{(f\ t+c)^n B\sigma_0^m\ e^{m\left(\frac{e^{kt}}{(f\ t+c)^n}+b\right)}}\right]^{\frac{1}{m-\beta}} \tag{3.4}$$

On Fig. 3.2, the theoretical damage curve for the case of a compressible medium according relation (3.4) is shown. In the calculations the following values of the coefficients were used: $B = 1.9 \cdot 10^{-23}\,[MPa]^{-8} \cdot [h]^{-1}$, $f = 0.08\,[h]^{-1}$, $b = 0$, $c = 10^4$, $k = 2.6 \cdot 10^{-5}\,[h]^{-1}$, $m = 8$, $n = 0.15$, $\beta = 2$, $\sigma_0 = 120\,[MPa]$.

3.4 Long-Term Strength Criteria

The criterion of long-term strength can be obtained under the condition when the damage parameter reaches a critical value. Taking in (4) the fracture condition in the form $t = t_f$, $\psi = \psi_*$, we can receive the following long-term strength criterion for

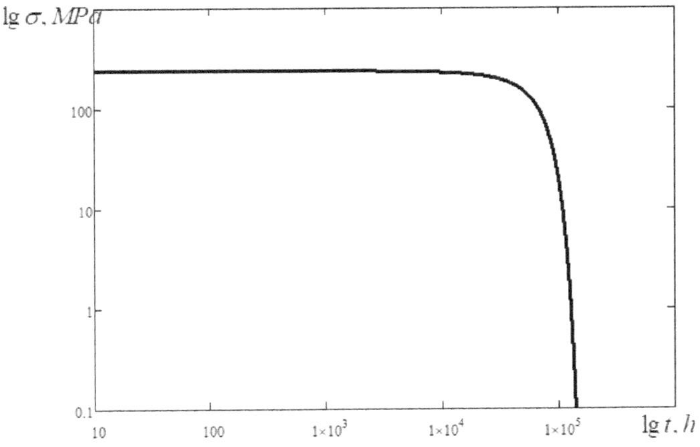

Fig. 3.3 The long-term strength curve according to relation (3.5)

the case of a compressible medium

$$\sigma = \left[\frac{e^{kt}\left(k - \frac{nf}{f\,t+c}\right)}{(f\,t+c)^n e^{m\left[\frac{e^{kt}}{(f\,t+c)^n}+b\right]} B\psi_*^{m-\beta}} \right]^{\frac{1}{m}} \tag{3.5}$$

On Fig. 3.3, the long-term strength curve according to relation (3.5) is shown. In the calculations the following values of the coefficients are used: $c = 10^4$, $b = 0$, $k = 2.6 \cdot 10^{-5}$ $[h]^{-1}$, $n = 0.15$, $f = 0.08$ $[h]^{-1}$, $m = 8$, $\beta = 2$, $B = 1.9 \cdot 10^{-23}$ $[MPa]^{-8} \cdot [h]^{-1}$, $\psi_* = 0.4$.

3.5 Conclusions

A method to determine the damage value using experimental creep curves is proposed. The compressibility of metallic materials is taken into account and the relative density changes are considered as a damage parameter. Only one kinetic equation for creep rate is formulated. From this equation the damage parameter is determined depending on the creep rate and the creep deformation. To describe the experimental creep curves, empirical dependences in the form of power, exponential and mixed functions are used. Theoretical continuity curves are plotted. The long-term strength criterion is obtained under the condition, when continuity parameter is reached the critical value. The corresponding theoretical long-term strength curves are constructed.

References

1. Kachanov LM (1958) About fracture time under creep conditions. Izv. USSR Academy of Sciences. OTN 8:26–31 (in Russian)
2. Rabotnov, YN (1959) On the mechanism of long-term fracture. Questions of strength of materials and structures. Izd-vo USSR Academy of Sciences 5–7 (in Russian)
3. Rabotnov YN (1966) Creep of structural elements. Science, Moscow (in Russian)
4. Arutyunyan RA (2017) The problem of high-temperature creep and long-term strength in the mechanics of materials. Dokl Phys 62(8):384–386
5. Arutyunyan A, Arutyunyan R, Saitova R (2019) The criterion of high-temperature creep of metals based on relative changes of density. WSEAS Trans Appl Theor Mech 14:140–144
6. Ratcliffe RT, Greenwood GW (1965) Mechanism of cavitation in magnesium during creep. Phil Mag 12:59–69
7. Betekhtin, VI (1997) Porosity of solids. Trans St-Petersburg Acad Sci for strength problems 1, 202–210
8. Arutyunyan AR, Arutyunyan RA, Saitova RR (2017) Evolution of the processes of damage to the elastic-plastic medium in high-temperature creep. In: Proceedings of the scientific and technical conference on the structural mechanics of the ship, dedicated to the memory of Professor Postnov VA and the 90th anniversary of his birth, St. Petersburg. 92–93 (in Russian)
9. Arutyunyan AR, Arutyunyan RA, Saitova RR (2020) High-temperature creep and damage of metallic materials. J Phys: Conf Ser 1474:012005
10. Arutyunyan AR, Saitova RR (2022) Exact and approximate solutions of the system of interrelated equations of the theory of creep and long-term strength. J Phys: Conf Ser 2231:012001
11. Arutyunyan AR, Arutyunyan RA., Saitova RR (2020) The definition of damage parameter changes from the experimental high-temperature creep curves. Lect Note Mech Engi 53–59
12. Hore S, Ghosh RN (2011) Computer simulation of the high temperature creep behavior of Cr-Mo steels. Mater Sci Eng A 528:6095–6102
13. Aghajani A (2009) Evolution of microstructure during long-term creep of a tempered martensite ferritic steel. Dissertation to obtain the degree of Doctor of engineer of the faculty of mechanical engineering of the Ruhr University Bochum. Bochum. 108
14. Panait CG et al (2010) Evolution of dislocation density, size of subgrains and MX-type precipitates in a P91 steel during creep and during thermal ageing at 600 ∘C for more than 100,000 h. Mater Sci Eng A 527:4062–4069

Chapter 4
Modeling of Thermoviscous Liquid Flow in a Conical Diffuser

Dilara R. Galeeva, Victor N. Kireev, and Said F. Urmancheev

Abstract In this paper, we study temperature effect on the laminar flow of an incompressible thermoviscous fluid in a conical diffuser. As a result of numerical simulation, velocity, temperature, and viscosity distributions in the diffuser channel were obtained, as well as the velocity profiles of the thermoviscous liquid flow for various sections. The influence of the channel geometry, temperature field and parameters of the dependence of viscosity on temperature on the hydrodynamic characteristics of the flow in a diffuser is studied.

Keywords Thermoviscous liquid · Diffuser · Mathematical modeling · Flow rate

4.1 Introduction

Diffuser is a section of pipeline, on which there is a smooth increase in the pipeline cross section. Diffusers are widely used in devices of chemical, oil, food and other industries to convert dynamic pressure into static pressure by slowing down the main flow rate. Various aspects of the flow in diffusers are studied in a huge number of theoretical, experimental and applied publications. In paper [1] it is proposed to use pipelines with a periodically changing cross section to improve the efficiency of pumping hydrocarbon liquids in transport systems. In practice, diffusers can be subjected to external thermal effects, which affect the flow parameters. The paper [2] proposes a refined approach to the calculation of the coefficient of local hydraulic resistance to the flow of a viscous fluid in a flat channel with a stepped narrowing under external thermal action.

In the present work, the effect of non-uniform temperature field on the laminar flow of an incompressible fluid with the temperature-dependent viscosity in a conical diffuser is studied.

D. R. Galeeva (✉) · V. N. Kireev
Ufa University of Science and Technology, Ufa, Russia
e-mail: lara_wood@mail.ru

S. F. Urmancheev
Mavlutov Institute of Mechanics, Ufa, Russia

H. Altenbach and V. Kuzkin (eds.), *Advanced Problems in Mechanics*,
https://doi.org/10.1007/978-3-031-80103-7_4

4.2 Problem Statement

We will consider the flow of an incompressible thermoviscous fluid in a diffuser under the action of a pressure drop Δp. Diffuser has length L, inlet section radius R_1, outlet section radius $R_2 > R_1$. The temperature of the incoming liquid is T_{in}, diffuser wall is maintained under different constant temperature $T_w < T_{in}$. Our goal is to determine the fluid velocity vector, pressure, temperature field and dynamic viscosity in the diffuser.

We will introduce a cylindrical coordinate system, the origin of which lies in the center of inlet section of the diffuser, the z axis coincides with the pipe axis (Fig. 4.1).

The mathematical model consists of the continuity equation, the Navier-Stokes equations, in which the dependence of viscosity on temperature is taken into account, and the energy conservation equation. The non-stationary equations of the mathematical model in axisymmetric formulation have form:

$$\frac{\partial(rv)}{\partial r} + \frac{\partial(ru)}{\partial z} = 0, \tag{4.1}$$

$$\rho\left(\frac{\partial v}{\partial t} + v\frac{\partial v}{\partial r} + u\frac{\partial v}{\partial z}\right) = -\frac{\partial p}{\partial r} + \frac{1}{r}\frac{\partial}{\partial r}\left(\mu(T)r\frac{\partial v}{\partial r}\right) + \frac{\partial}{\partial z}\left(\mu(T)\frac{\partial v}{\partial z}\right) - \frac{\mu(T)v}{r^2}, \tag{4.2}$$

$$\rho\left(\frac{\partial u}{\partial t} + v\frac{\partial u}{\partial r} + u\frac{\partial u}{\partial z}\right) = -\frac{\partial p}{\partial z} + \frac{1}{r}\frac{\partial}{\partial r}\left(\mu(T)r\frac{\partial u}{\partial r}\right) + \frac{\partial}{\partial z}\left(\mu(T)\frac{\partial u}{\partial z}\right), \tag{4.3}$$

$$\frac{\partial T}{\partial t} + v\frac{\partial T}{\partial r} + u\frac{\partial T}{\partial z} = \alpha\left(\frac{1}{r}\frac{\partial}{\partial r}\left(r\frac{\partial T}{\partial r}\right) + \frac{\partial^2 T}{\partial z^2}\right), \tag{4.4}$$

where v and u are radial and axial velocity components, p is pressure, T is temperature, t is time, α is the thermal diffusivity, ρ is the density, $\mu(T)$ is the temperature dependence of dynamic viscosity of the liquid.

The boundary conditions are written as follows:
inlet ($z = 0$): $p = P_{in},\ T = T_{in}$,
outlet ($z = L$): $p = P_{out}$,
on the wall: $v = u = 0,\ T = T_w$.

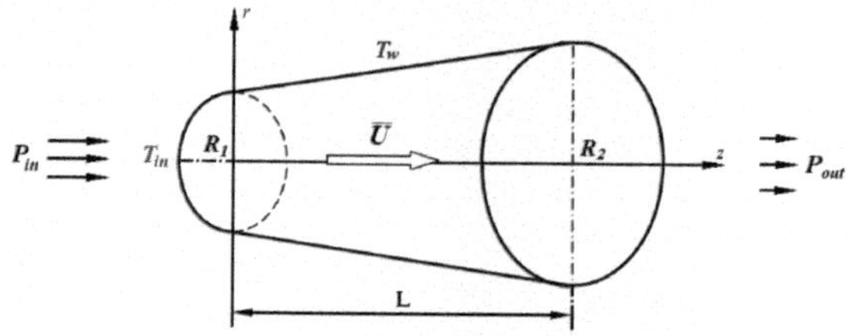

Fig. 4.1 Scheme of a conical diffuser

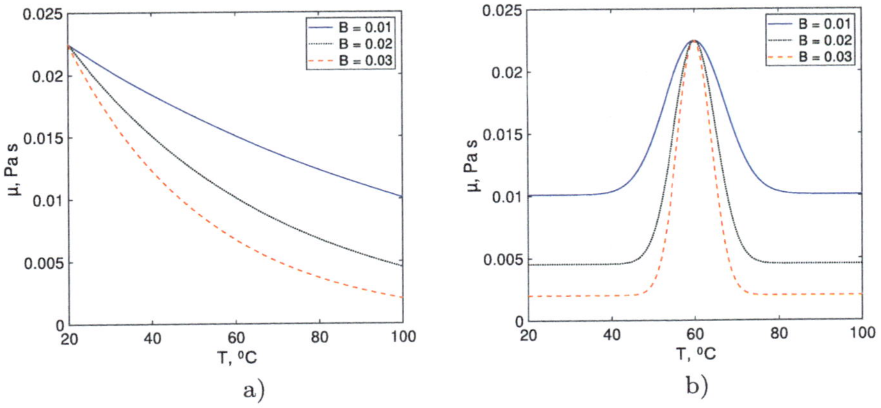

Fig. 4.2 Monotonic (**a**) and non-monotonic (**b**) dependences of viscosity on temperature

In this paper, we consider two types of viscosity [3]:
(1) monotonic dependence of viscosity on temperature (the viscosity of a liquid depends on temperature according to an exponential law):

$$\mu(T) = \mu_{max}\, e^{-B(T-T_w)}, \tag{4.5}$$

(2) non-monotonic dependence of viscosity on temperature:

$$\mu(T) = \mu_{min}\left[1 + A\, e^{-B(T-T_*)^2}\right], \tag{4.6}$$

where $B > 0$ is a parameter describing the nature of the change in viscosity,
$T_* = 0.5\,(T_{in} + T_w)$ – average temperature,
$A = e^{-B(T_w - T_{in})} - 1$ is the liquid anomaly parameter.

Figure 4.2 shows monotonic and non-monotonic dependences of viscosity on temperature for various parameters B. The larger parameter B, the stronger viscosity change. The case with $B = 0$ corresponds to a constant viscosity.

4.3 Numerical Results

The problem was solved numerically by the control volume method using the SIMPLE algorithm [4], modified to take into account the variable viscosity coefficient.

As a preliminary study, we considered the problem of flow in a circular cylinder, the exact solution of which is known [5]. Pipe length $L = 0.1$ m, radius $R = 0.01$ m. Liquid is a model fluid (70% solution of glycerin in water). The dynamic viscosity is set constant $\mu = 0.0184$ Pa s. The temperature is the same everywhere $T_{in} = T_w =$

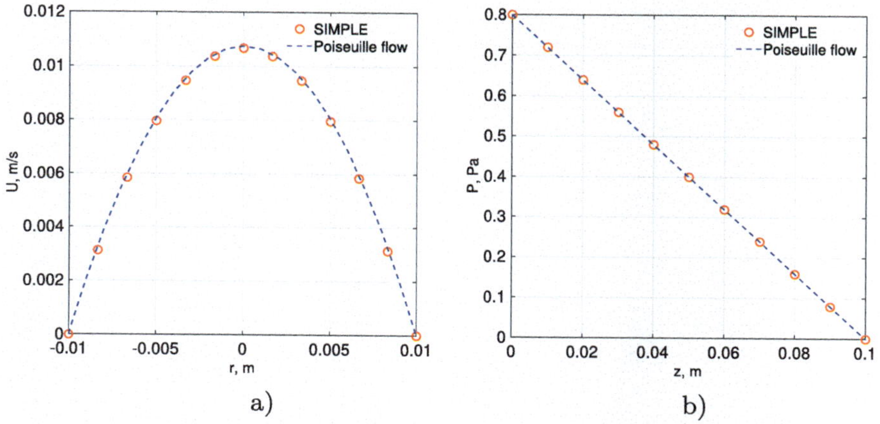

Fig. 4.3 Numerical and analytical solutions for axial velocity component u (**a**) and pressure p (**b**) in a cylindrical pipe

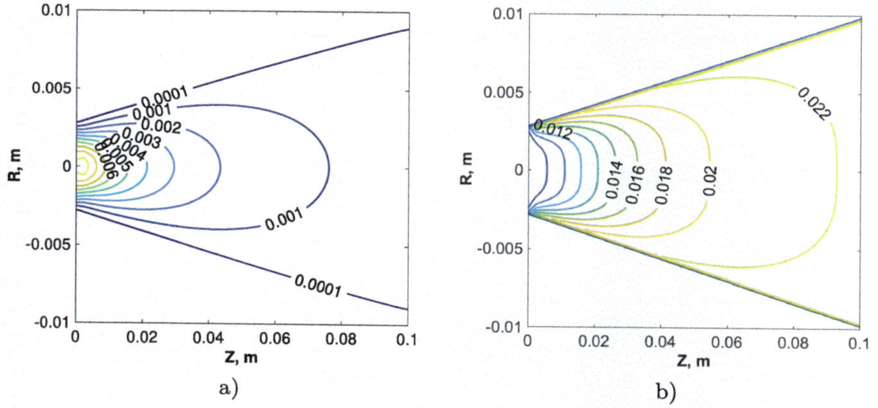

Fig. 4.4 Axial velocity component u (**a**) and dynamic viscosity $\mu(T)$ (**b**) for monotonic viscosity function in diffuser

40 °C. The exact and numerical solutions are compared. The graphs in Fig. 4.3 show that the numerical and analytical solutions are the same.

Now consider the fluid flow in a diffuser. We calculate flow of the same model fluid. Diffuser length $L = 0.1$ m, $R_2 = 0.01$m, $R_1 = 0.3R_2$. Inlet temperature $T_{in} = 100$ °C, wall temperature $T_w = 20$ °C, pressure drop $\Delta p = 0.8$ Pa, parameter describing the nature of the change in viscosity $B = 0.02$.

Figure 4.4 shows axial velocity component u and viscosity $\mu(T)$ for **monotonic viscosity function**. It can be seen that by the end of the diffuser the liquid has cooled down and its velocity has dropped. The viscosity graph corresponds to temperature and speed, as the end of the diffuser is approached, the viscosity increases.

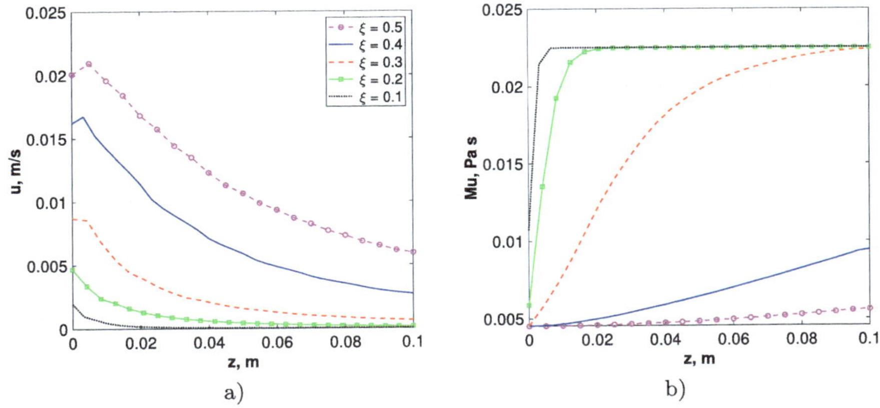

Fig. 4.5 Axial velocity u (**a**) and monotonic viscosity $\mu(T)$ (**b**) distribution for diffusers with different $\xi \in [0.1; 0.5]$

Geometry dependency. Let's introduce a parameter showing how many times the inlet radius is less than the outlet radius: $\xi = R_1/R_2$. The entry radius will change from half the exit radius ($\xi = 0.5$) to one tenth ($\xi = 0.1$). Consider how changing this parameter affects the velocity and viscosity.

The graphs in Fig. 4.5 show how parameters change as diffuser inlet radius decreases. As the inlet radius decreases, fluid velocity decreases. Temperature stops rising, liquid does not warm up. For diffusers with a small inlet radius, the viscosity quickly becomes constant $\mu = 0.0225$ Pa s, this value corresponds to a temperature 20 °C. It means that the larger opening angle of diffuser, the faster liquid cools.

Now consider the **non-monotonic viscosity function**. Viscosity on diffuser walls and viscosity of the hot inflowing fluid are the same (see Fig. 4.2b). Figure 4.6 shows axial velocity component and viscosity for this case. Viscosity graph changed the most. The yellow bands are high viscosity areas. There is a slowdown in the flow.

Consider how changing parameter ξ affects the velocity and viscosity. As in the monotonic case, the smaller inlet radius, the lower flow velocity. But its behavior changes. Figure 4.7a shows that velocity curves are not so smooth as in monotonic case (see Fig. 4.5a), they curve sharply. The high viscosity areas (Fig. 4.7b) slow down flow velocity. The liquid flows more slowly not only due to expansion of the diffuser, but also due to the viscous barrier. So liquid has time to cool down.

Areas of high viscosity also affect the velocity profile. Now it is not parabolic, in contrast to the monotonic case (see Fig. 4.8).

Finally, let's consider the volume flow rate Q. Figure 4.9 compares flow rate for monotonic, non-monotonic and constant viscosity (corresponds to $T = 100\,°\text{C}$). We calculated the flow rate for different inlet diameters, that is, for different ξ. It can be seen that the flow rate is greatest when the liquid is constantly hot. In the case of

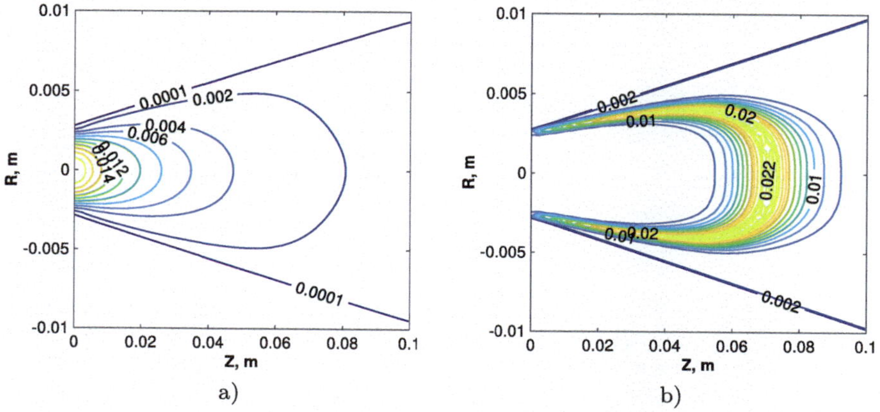

Fig. 4.6 Axial velocity component u (**a**) and dynamic viscosity $\mu(T)$ (**b**) for non-monotonic viscosity function in diffuser with $\xi = 0.3$

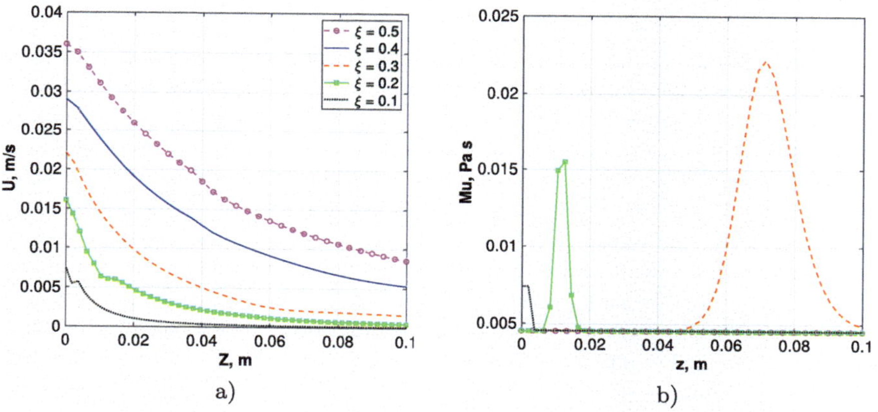

Fig. 4.7 Axial velocity u (**a**) and non-monotonic viscosity $\mu(T)$ (**b**) distribution for diffusers with different $\xi \in [0.1; 0.5]$

monotonic viscosity, the liquid in the diffuser cools, its velocity drops, and hence the volume flow rate of the liquid also decreases. For the nonmonotonic case at small ξ, flow rate almost coincides flow rate with a constant viscosity, because the viscosity of the cold liquid is the same as that of the hot liquid. But with an increase ξ, the flow rate decreases due to the slowing down of the liquid due to areas of high viscosity.

From a geometric point of view, it can be seen that the larger opening angle of the diffuser (the smaller ξ), the lower flow rate Q.

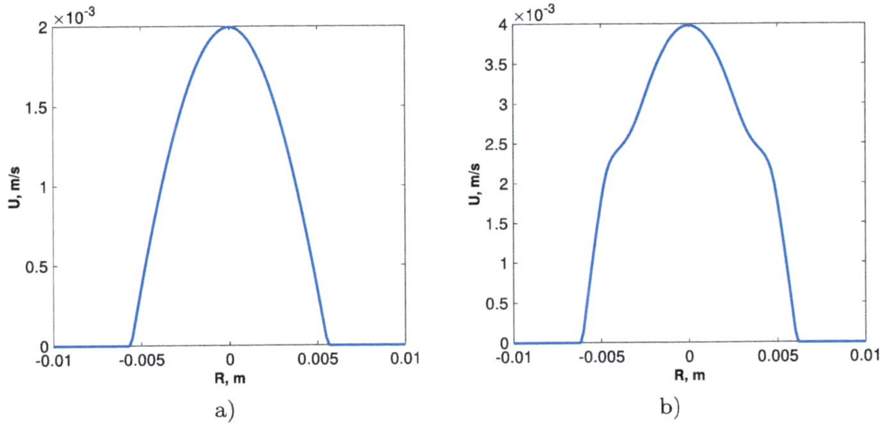

Fig. 4.8 Axial velocity profile u for monotonic (**a**) and non-monotonic (**b**) viscosity functions in diffuser with $\xi = 0.3$

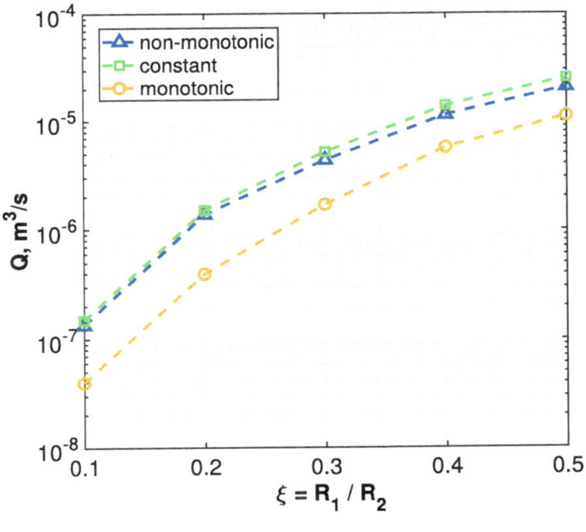

Fig. 4.9 Volume flow rate Q for diffusers with different ξ for 3 types of viscosity

As a result of numerical simulation, the velocity, temperature, and viscosity fields in the diffuser channel are obtained, as well as the velocity profiles of the thermoviscous liquid flow for various sections. The influence of the channel geometry, temperature field, and parameters of the temperature dependence of viscosity on the hydrodynamic characteristics of the flow in a diffuser is studied.

Acknowledgements The research is supported by the Russian Science Foundation (Project No. 22-21-00915).

References

1. Khabibullin MYa (2019) Study of liquid flow processes in pipeline with changing diameter. Problemy sbora, podgotovki i transporta nefti i nefteproduktov (1):89–95. [in Russian]
2. Nizamova AD, Kireev VN, Urmancheev SF (2021) To the calculation of the channel local hydraulic resistance with external heating. Vestnik UGATU 25(3):34–41. [in Russian]
3. Urmancheev SF, Kireev VN (2004) Steady flow of a liquid with a temperature anomaly of viscosity. Doklady Phys 49:328–331
4. Patankar S (1980) Numerical heat transfer and fluid flow. Hemishere Publishing Corporation, New York, p 199
5. Slezkin NA (1955) Dynamics of a viscous incompressible fluid. Moscow: Gostekhizdat 520 p [in Russian]

Chapter 5
Time Optimization of Constrained Control for a Thermoelectric Solid System with a Peltier Element

Alexander Gavrikov and Georgy Kostin

Abstract A solid system consisting of two heat conducting cylinders with a thermoelectric converter (Peltier element) between them is considered. A nonlinear model, which was previously verified by authors, is used to design a constrained control law that allows us to achieve a steady-state distribution of the temperature in one of the cylinders in considerably less time than the characteristic time of transient processes. The initial-boundary value problem is exactly linearized over temperature by means of feedback linearization. Although the resulting system is nonlinear in a control function, it is possible to construct a finite-dimensional approximation based on analytical solution of the corresponding eigenproblem for a constant control signal. The time-optimal control problem is studied numerically by using this eigenfunction decomposition. To construct admissible control laws, an auxiliary unconstrained optimization problem is introduced. Its cost functional represents a weighted sum of temperature deviation from the desired zero distribution and a penalty for violating an electric power constraint. The control time interval is split into several parts, and on each subinterval the control signal is taken constant. The optimal piecewise constant feedforward control is found numerically by applying the gradient descent method. We analyze the proposed control law with respect to the shortest admissible time of the process.

Keywords Time-optimal control · Constrained optimization · Thermoelectric solid system · Peltier element

A. Gavrikov (✉) · G. Kostin
Ishlinsky Institute for Problems in Mechanics RAS, Moscow, Russia
e-mail: gavrikov@ipmnet.ru; Alexander.Gavrikov@psu.edu
URL: https://sites.psu.edu/alexandergavrikov/

G. Kostin
e-mail: kostin@ipmnet.ru

A. Gavrikov
The Pennsylvania State University, University Park, PA 16802, USA

H. Altenbach and V. Kuzkin (eds.), *Advanced Problems in Mechanics*,
https://doi.org/10.1007/978-3-031-80103-7_5

5.1 Introduction

At present, solid-state thermoelectric converters, including Peltier elements (PEs), are increasingly used in technical applications. Despite moderate power of such devices, they have a number of significant advantages, such as compactness, resistance to external disturbances, accuracy and rapid response rate of control signals. Applying thermoelectric converters as actuators and sensors makes it possible to create efficient heat transfer systems and control various technological processes to achieve and maintain certain temperature regimes in solids. The examples of such application, including systems utilizing renewable energy sources, can be found in farming [1], machining [2], energy generation [3], air conditioning [4], water heating [5] and cooling [6]. For the control of such systems, a wide variety of methods is used from bang-bang type control [7] to classical PI controllers as well as neural network control models [8].

Often, a thermoelectric system is modeled via a lumped model [7], which simplifies an optimal control problem and makes it possible to find an analytical solution. We consider it as a system with distributed parameters and are interested in the spatial distribution of the temperature, which makes the control problem much harder.

To construct accurate control laws, it is necessary to use coupled nonlinear thermoelectric models that take into account effects of electric energy recuperation, Joule heat losses, and heat exchange with the environment. Given limitations imposed on the power consumption of a PE, it is important to evaluate both reachability ranges for admissible input signals and the minimal time in which the system can be transfered to the desired state. Even for relatively simple systems, a direct solution of the time-optimal control problem appears to be a rather challenging goal. Therefore, one can use different approaches to numerically estimate the minimum of control time. In this paper, we apply the gradient descent method for this purpose. For a system consisting of two metal cylinders and a Peltier disk-like element separating them, we aim to reach a small vicinity of the desired terminal state, while the constraints on the electric current are taken into account utilizing a penalty term added to the cost functional.

This control strategy could be supplemented by additional feedback to compensate for disturbances, as in [9], or by implementing a model predictive controller [8] that maintains the temperature around the level prescribed by the feedforward control strategy.

5.2 Model of a Thermoelectric Converter

A full controlled plant may contain many actuated elements consisting of parts with complex geometry and inhomogeneous parameters, what hinders a study of limiting behavior. For this reason, as a simple example we consider a thermoelectric system consisting of two identical heat conducting cylinders and a thin circular PE between

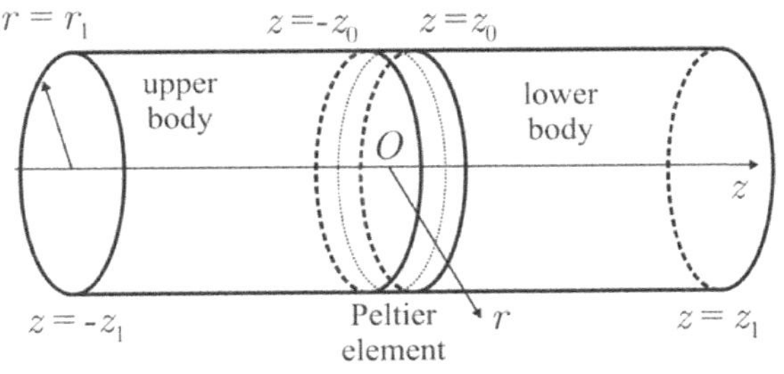

Fig. 5.1 Schematic representation of the experimental setup

them, see Fig. 5.1. The length of the cylinders and the PE are $z_1 - z_0$ and $2z_0$, respectively, where $z_0 \ll z_1$. Their radii are equal to r_1. A corresponding experimental setup was constructed at the Chair of Mechatronics of the University of Rostock, Germany [10, 11].

Under assumption that the PE characteristics are constant and heat transfer in the PE occurs only in one direction, a nonlinear model of thermoelectric processes was proposed and validated in [10–13]. The governing equations of this model were simplified by introducing a feedback linearization control in [14]. The resulting initial-boundary value problem (IBVP) has the following form in cylindrical coordinates $\mathbf{x} = (r, \phi, z)$

$$
\begin{gathered}
c_p \rho_p \dot{\theta} = \lambda_p \theta''_{zz} + \tfrac{u^2}{R|\mathcal{V}_p|}, \quad \mathbf{x} \in \mathcal{V}_p, \\
c_a \rho_a \dot{\theta} = \lambda_a \Delta\theta, \quad \mathbf{x} \in \mathcal{V}_k, \quad k = 1, 2, \\
\theta'_z|_{|z|=z_1} = 0, \quad \alpha\theta + \lambda_a \theta'_r|_{r=r_1,|z|>z_0} = \alpha\theta_A, \\
\theta|_{|z|=z_0 \pm 0} = \theta|_{|z|=z_0 \mp 0}, \\
-\lambda_a \theta'_z\big|_{|z|=z_0+0} = \left[(\theta + \theta^0)\tfrac{Su}{R|\mathcal{A}_p|} - \lambda_p \theta'_z\right]_{|z|=z_0-0}, \\
\theta(0, \mathbf{x}) = \Theta(\mathbf{x}), \\
u(t) = \begin{cases} u_0(t) - u_- & \text{for} \quad u_0 < u_- \\ 0 & \text{for} \quad u_- \le u_0 \le u_+ \, . \\ u_0(t) - u_+ & \text{for} \quad u_0 > u_+ \end{cases}
\end{gathered}
\tag{5.1}
$$

Here, $c_{p,a}$ are specific heat capacities, $\rho_{p,a}$ are densities, $\lambda_{p,a}$ are thermal conductivities for the PE (p) and the cylinders (a), α is the heat transfer coefficient, θ_A and θ^0 are the constant ambient and reference temperatures, respectively, $|\mathcal{A}_p| = \pi r_1^2$, S is the Seebeck coefficient, and R is the ohmic resistance of the PE. The cylinders and the PE occupy domains $\mathcal{V}_k = I_r \times I_\phi \times I_k$ ($k = 1, 2$) and $\mathcal{V}_p = I_r \times I_\phi \times I_p$, respectively, where $I_r = [0, r_1]$, $I_\phi = [0, 2\pi]$, $I_p = [-z_0, z_0]$, $I_1 = [z_0, z_1]$, $I_2 = [-z_1, -z_0]$. The temperature distribution $\theta(t, \mathbf{x})$ and the ambient temperature are measured relatively to the reference temperature θ^0. Although we suppose that θ_A is constant, the proposed approached can be extended to the ambient temperature varying in time [9]. For definiteness, it is assumed that the thermoelectric system (5.1) is initially in a steady state: $\Theta(\mathbf{x}) = \theta_{st}$, such that the cylinder $\mathcal{V}_1$ has average temperature θ_{av}, and u_{st} is a constant voltage that yields the temperature distribution θ_{st}.

The control function $u(t)$ is defined through the input voltage $u_0(t)$ and thresholds $u_\pm$. The total control voltage supplied to the PE is expressed as

$$U = u_0 + S[\tilde{\theta}]. \tag{5.2}$$

Here, $[\tilde{\theta}]$ denotes the jump of the average temperature on the top and bottom sides of the PE. Although the voltage $S[\tilde{\theta}]$ provides the linearization w.r.t. the temperature, see [14], the resulting IBVP (5.1) is still nonlinear in u. Therefore, an explicit design of optimal control is hindered. In what follows, we propose a semi-analytical approach to solution of an optimal control problem (OCP) for (5.1) based on eigenfunction decomposition of θ for constant values of u.

5.3 Optimal Control Problem

Previously, we considered an OCP for the IBVP (5.1), where the goal was to achieve a steady-state distribution of the temperature in one of the cylinders. Although both cylinders are actuated by the PE in such a problem, the second cylinder $\mathcal{V}_2$ serves only as a "heat sink", while the first one, $\mathcal{V}_1$, represents a controlled part of a plant in which a stable working regime has to be achieved. To this end, we utilized a piecewise control function [14] and took into account disturbances of the ambient temperature [9]. Recently, we optimized a constrained piecewise control signal [15].

5.3.1 Control Problem: General Statement

In this work, we consider a time-optimal control problem. We suppose that the goal is to return the actuated cylinder to the zero state in a shortest time T :

$$T \to \min_u \text{ subject to } \frac{\|\theta(T, \mathbf{x})\|_{L_2(\mathcal{V}_1)}}{\|\theta(0, \mathbf{x})\|_{L_2(\mathcal{V}_1)}} \leqslant \varepsilon \tag{5.3}$$

as well as the IBVP (5.1) and constraints on u:

$$|u| \leq |u_{st}|. \tag{5.4}$$

That is, we suppose that the bounds on u are defined by a control signal u_{st} corresponding to the initial steady state θ_{st}.

5.3.2 Relaxed Control Problem

Direct solution of (5.3), (5.4) is hindered due to its nonlinearity and uncertainty of existence of global or even local minimum. To deal with these difficulties, we consider a relaxed version of (5.3), (5.4) instead of finding exact solution. We find admissible control laws such that $\frac{\|\theta(T,\mathbf{x})\|_{L_2(\mathcal{V}_1)}}{\|\theta(0,\mathbf{x})\|_{L_2(\mathcal{V}_1)}} \leqslant \varepsilon$ and (5.1), (5.4) are satisfied. That allows for estimating from above the minimal control time T_{min}. To design numerically admissible control laws, we consider an unconstrained optimization problem keeping only (5.1) as a constraint. To this end, we minimize the cost function

$$F \to \min_u, \quad F = 10^{\gamma} F_d + F_p, \quad F_d = c_\theta \|\theta(T,\mathbf{x})\|^2_{L_2(\mathcal{V}_1)},$$
$$c_\theta = (|\mathcal{V}_1|\theta^2_{av})^{-1} 2\pi \int_0^{r_1} r J_0^2(\mu_{0,0} r/r_1) dr, \tag{5.5}$$

subject to the IBVP (5.1). In (5.5), J_0 is the Bessel function of zeroth order of the first kind, $\mu_{0,0}$ is a root of an equation presented in the next section. Note that the L_2-norm in (5.5) is evaluated only over $\mathcal{V}_1$. That is, we minimize the temperature only in the controlled cylinder $\mathcal{V}_1$, while the actuated cylinder $\mathcal{V}_2$ (heat sink) may have any temperature yielded by u.

The functional F_d in (5.5) represents the cost of the deviation of the temperature distribution from the zero state in the controlled cylinder $\mathcal{V}_1$ at time T, while

$$F_p = c_u \int_0^T f(u) dt, \quad c_u = (\rho_a c_a R |\mathcal{V}_1| \theta_{av})^{-1}, \tag{5.6}$$

is a penalty term.

The penalty functional F_p was proposed in [15] and characterized as follows:

$$\begin{aligned} f(u) \sim o(u^2) \text{ for } |u| < \varepsilon, \quad \varepsilon \ll u_{st}, \\ f(u) \ll u^2 \quad \text{for} \quad |u| < u_{st} - \varepsilon, \\ f(u) \gg u^2 \quad \text{if} \quad |u| \geq u_{st}. \end{aligned} \tag{5.7}$$

Therefore, F_p is small enough if the control function satisfies the constraints and is large otherwise. We define the penalty function according to

$$f(u) = c_1 (e^{c_2 u^2} - c_2 u^2 - 1). \tag{5.8}$$

The constants c_1, c_2 must be chosen such that the conditions (5.7) are satisfied. See Fig. 5.2, where these constants are as in the numerical example presented further.

To estimate the minimal control time T_{min} in (5.3), we solve the problem (5.5) for an admissible control u via the gradient descend method for T in some range $[T_0, T_1]$ and choose as T_{min} such a value that the constraints (5.4) are satisfied.

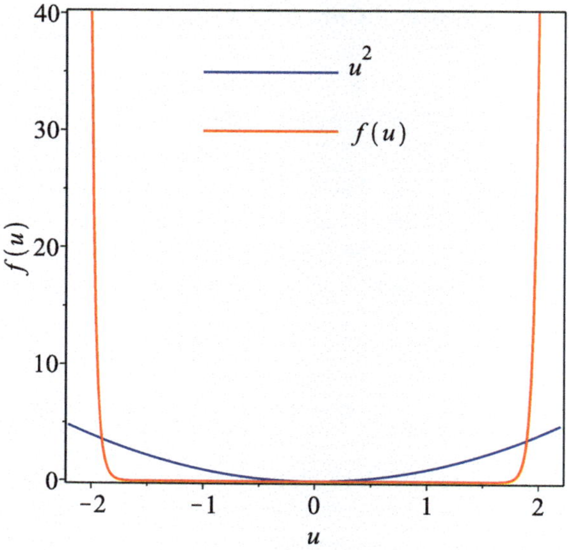

Fig. 5.2 Penalty function $f(u)$ vs u^2

5.4 Finite Dimensional Approximation

To simplify the studied control problem, we consider a finite dimensional approximation of the IBVP (5.1). Suppose that $u(t)$ is constant: $u(t) = u^{(i)}$. Then a solution to (5.1) can be represented as the series

$$\begin{aligned} \theta &= \sum\nolimits_{n,m,k} e^{\nu_{n,m,k}t}\,\Xi_{n,m,k}(r,\phi,z), \\ \Xi_{n,m,k}(r,\phi,z) &= J_n(\mu_{n,m}r/r_1)\cos(n\phi)\psi_k(z), \end{aligned} \tag{5.9}$$

where Ξ are orthogonal eigenfunctions of the IBVP (5.1). In (5.9), J_n are Bessel functions of the first kind of order n, $\mu = \mu_{n,m}$ are roots of the equation

$$\alpha J_n(\mu) + \frac{\lambda_a}{r_1}(nJ_n(\mu) - \mu J_{n+1}(\mu)) = 0, \tag{5.10}$$

which corresponds to the boundary condition on the lateral surface of the cylinders. The quantities

$$\nu = \tau^{-1} = -\frac{\lambda_a(\mu^2 + \xi^2)}{c_a\rho_a r_1^2}$$

are the inversed characteristic times of decay $\tau = \tau_{n,m,k}$ of each mode, and $\xi = \xi_k$ are eigenvalues corresponding to the eigenfunctions in the z-direction $\psi(z) = \psi_k(z)$ [14]. These functions are found explicitly as piecewise continuous linear combinations of exponential and trigonometric functions.

Analysis of characteristic times $\tau_{n,m,k}$ allows one to determine the modes that can be neglected on the time interval $[0, T]$ due to their rapid decay. For the experimental setup considered here, only the four lowest modes have characteristic times $\tau_{n,m,k} \gtrsim$ 10 sec. For the highest modes, $\tau_{n,m,k} \lesssim 5$ sec, see [14]. Therefore, these transient processes can be excluded from consideration if the intervals of constancy of u are much longer. It is worth mentioning that the angular modes ($n > 0$) are uncontrollable in this setting. However, they belong to the group of quickly decaying modes and can be neglected.

Hence, the expansion (5.9) simplifies to

$$\theta = \sum_{k=0}^{3} e^{\nu_{0,0,k} t} J_n(\mu_{0,0} r / r_1) \psi_k(z). \tag{5.11}$$

The expansion of the initial steady state $\Theta(\mathbf{x}) = \theta_{st}$ w.r.t. to these eigenfunctions yields a vector of coefficients $\bar{\Theta}$. Next, by substituting (5.11) into (5.1) and projecting the result onto the eigenfunctions, we obtain the four-dimensional system of ordinary differential equations

$$\begin{aligned} \dot{\bar{\theta}} &= A\bar{\theta} + G(u, \theta_A, \theta^0), \quad \bar{\theta}(0, z) = \bar{\Theta}, \\ \bar{\theta} &= (\theta_0, \ldots, \theta_3)^T, \quad \bar{\Theta} = (\Theta_0, \ldots, \Theta_3)^T, \\ \Theta_i &= \int_{\mathcal{V}} \Theta(\bar{x}) \Xi_i(\mathbf{x}) d\mathbf{x}, \quad \Xi_i = J_0\left(\mu_{0,0} r / r_1\right) \psi_{0,0,i}. \end{aligned} \tag{5.12}$$

Here, $\theta_i = \theta_{0,0,i}$, the matrix A is diagonal formed by the eigenvalues $\nu_{0,0,i}$, and the vector G is constant for a constant u. Hence, the system (5.12) can be solved explicitly for any given initial conditions (ICs) $\bar{\Theta}$.

The steady state θ_{st}, which we take as ICs in (5.1), is obtained for a given voltage u_{st} in a similar way. Neglecting the rate of temperature change in (5.1), we plug into the resulting system the ansatz function $\theta_{st}(\mathbf{x}) = J_0(\mu_{0,0} r / r_1) \psi_{st}(z)$ and find ψ_{st} as a function of u_{st} explicitly as a piecewise continuous linear combination of cosines and quadratic polynomials.

In its turn, the voltage u_{st} is chosen such that the resulting temperature distribution θ_{st} has some prescribed average value θ_{av} in the actuating cylinder: $\theta_{av} = \frac{1}{|\mathcal{V}_1|} \int_{\mathcal{V}_1} \theta_{st} dV$. Utilizing the procedure described above and functions $\psi_{st}(u_{st})$, we find u_{st} explicitly by solving the minimization problem

$$\left(\frac{1}{|\mathcal{V}_1|} \int_{\mathcal{V}_1} \theta_{st} dV - \theta_{av} \right)^2 \to \min_u . \tag{5.13}$$

5.5 Solution of the Control Problem

5.5.1 Auxiliary Optimization Problem

In this section, a combined numerical-analytical approach to the solution of OCP (5.3) is proposed assuming that the control function $u(t)$ is piecewise constant.

For a given T, the time interval $[0, T]$ is split into 5 equal subintervals $[t_i, t_{i+1}]$, $t_0 = 0$, $t_5 = T$, $i = 0, \ldots, 5$. We suppose that $u(t)$ is constant on each subinterval: $u(t) = u^{(i)} \in \mathbb{R}$ for $t \in [t_i, t_{i+1}]$. For a given $u^{(i)}$, the eigenfunctions $\psi_k^{(i)}(z)$ are found according to Sect. 5.4. Next, we solve explicitly (5.12) on $[0, t_1]$ for the given initial steady state $\bar{\Theta}$ and the control function $u(t) = u^{(0)}$. The resulting distribution $\theta(t_1, \mathbf{x})$, which is a linear combination of eigenfunctions $\psi_k^{(0)}(z)$, is re-expanded w.r.t. the eigenfunctions $\psi_k^{(1)}(z)$ corresponding to $u^{(1)}$. Taking coefficients of this expansion as ICs $\bar{\Theta}^{(1)}$, we solve on $[t_1, t_2]$ the system (5.12) with a new matrix $A^{(1)}$ determined by eigenvalues $\nu_k^{(1)}$. This yields the temperature $\theta(t_2, \mathbf{x})$. We repeat consequently solution of (5.12) and re-expansion of θ until the terminal state $\theta(T, \mathbf{x})$ is reached.

For the fixed time horizon T, we find piecewise constant $u = (u^{(0)}, \ldots, u^{(4)})^T$ such that

$$F[u, \theta] \to \min_{u(t), t \in [0,T]}, \tag{5.14}$$

where F is defined in (5.5). This problem is solved via the gradient descend. When starting with a vector

$$\bar{u}^{(0)} = (u^{(0,0)}, \ldots, u^{(0,4)}), \tag{5.15}$$

the next value is computed according to

$$\bar{u}^{(j+1)} = \bar{u}^{(j)} + \varepsilon \nabla_{\bar{u}} F, \quad \varepsilon \ll 1. \tag{5.16}$$

The gradient $\nabla_{\bar{u}} F$ is found numerically via finite differences

$$\frac{F(u^{(j,l)} + \delta u^{(j,l)}) - F(u^{(j,l)})}{\delta}, \quad \delta \ll 1, \tag{5.17}$$

while the values of $F(u^{(j,l)})$ and their variation $F(u^{(j,l)} + \delta u^{(j,l)})$ are obtained analytically by using the algorithm presented in Sect. 5.4.

5.5.2 Estimating the Minimal Time

To estimate the minimal time T_{min}, we start with $T_0 \sim \tau_{0,0,0}$, where $\tau_{0,0,0}$ is the characteristic time of decay of the zeroth eigenmode. For such large T, the system loses most of the heat due to natural cooling/heating and the active control is hardly required. Next, we decrease T, $T_j = T_{j-1} - \delta T_0$ and solve the OCP (5.14) on the time interval $[0, T_j]$. The minimal value of T_j such that the solution $u(t)$ of (5.14) does not violate the constraints, we take as an estimate from above of T_{min}.

In the next section, we present a numerical example of implementation of approaches described in Sects. 5.4, 5.5.

5.6 Numerical Results and Discussion

The following physical parameters corresponding to the experimental setup, see [10, 11], are considered: $\lambda_a = 254.4$ W/m/K, $\lambda_p = 0.517$ W/m/K, $\rho_a = 2700$ kg/m^3, $\rho_p = 3000$ kg/m^3, $c_a = 896$ J/kg/K, $c_p = 500$ J/kg/K, $\alpha = 8.4$ W/m^2/K, $h = 0.1$ m, $r_1 = 0.031$ m, $z_0 = 0.00195$ m, $z_1 = z_0 + h$, $S = 0.0427$ W/K/A, $R = 6.03\ \Omega$, $u_+ = 1.115$ V and $u_- = -1.29$ V. We take the reference temperature $\theta^0 = 293$ K, and the initial steady state θ_{st} such that the average temperature θ_{av} in the controlled cylinder $\mathcal{V}_1$ is 5.5 K. The voltage u_{st} corresponding to this state is found solving (5.13): $u_{st} \approx 1.44$ V. We take this value as the constraints: $|u(t)| \leq 1.44$ V. The terminal time T is varied with the step $\delta T_0 = 12$ sec.

To design admissible control laws, we consider several values of the weighting coefficient γ in (5.5). The larger value of γ favors minimization of the temperature rather than satisfaction of the control constraints. In Figs. 5.3, 5.4, 5.5, 5.6, 5.7, 5.8 and 5.9 the numerical results for the values of the weighting coefficient $\gamma = 3, 4, 5$ are presented. In Fig. 5.3, the optimal values of the cost function F (in logarithmic scale) are shown as a function of the time horizon T. These values do not change significantly while $T \gg 10^3$ sec. In this region, the obtained control laws satisfy constraints and transfer the system to a small vicinity of the zero state.

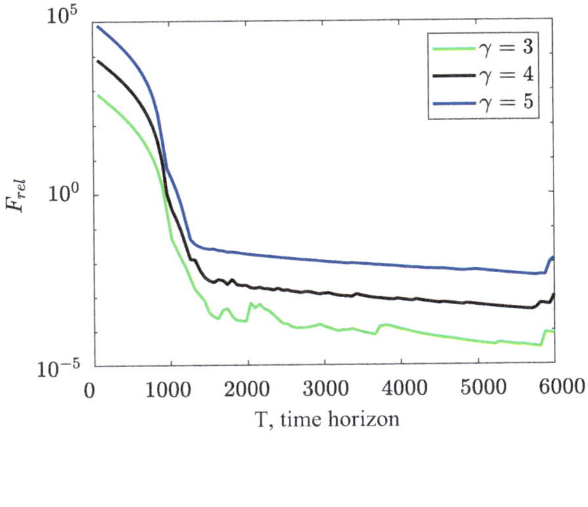

Fig. 5.3 The relative value of the cost functional F

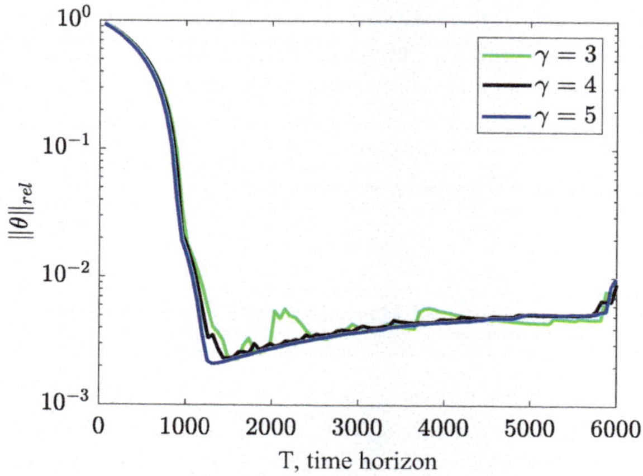

Fig. 5.4 The norm of the terminal temperature distribution in the controlled cylinder related to the norm of temperature at $t = T$ due to natural cooling

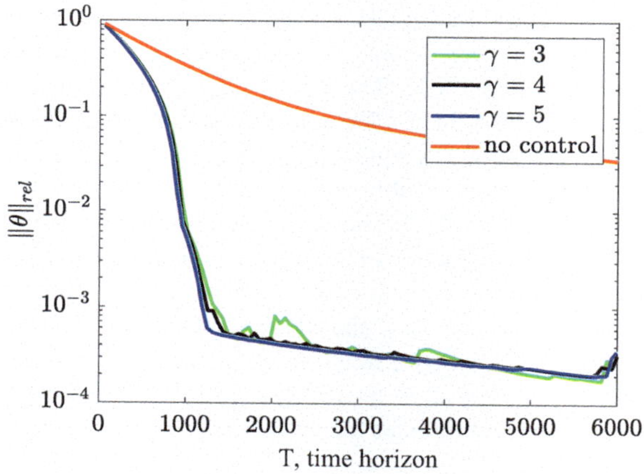

Fig. 5.5 The norm of the terminal temperature distribution in the controlled cylinder related to the norm of initial temperature distribution

This statement is supported by Figs. 5.4, 5.5, where the relative L_2-norm of the terminal temperature distribution in $\mathcal{V}_1$ is given in logarithmic scale. In Fig. 5.4, these values are normalized by the norm of the temperature in $\mathcal{V}_1$ in this cylinder resulting from natural cooling during the same time T. Therefore, Fig. 5.4 shows the effectiveness of the control law. As it can be expected, this effectiveness grows as T decreases: on long time intervals the natural cooling is enough to achieve a state close to zero. However, as T becomes smaller, the advantage of an active control is more clear: it allows to make temperature up to 3 order smaller than the natural cooling. Nevertheless, since power required to achieve the zero state grows with decreasing T, for $T \gtrsim 10^3$ the control law within the constraints loses its advantages and the effectiveness diminishes quickly. Figure 5.4 demonstrates the temperature decreasing comparing with the initial steady state. Here, the L_2-norm of the terminal temperature distribution in $\mathcal{V}_1$ (in logarithmic scale) is related to the L_2-norm of the steady state. The red curve represents the natural cooling.

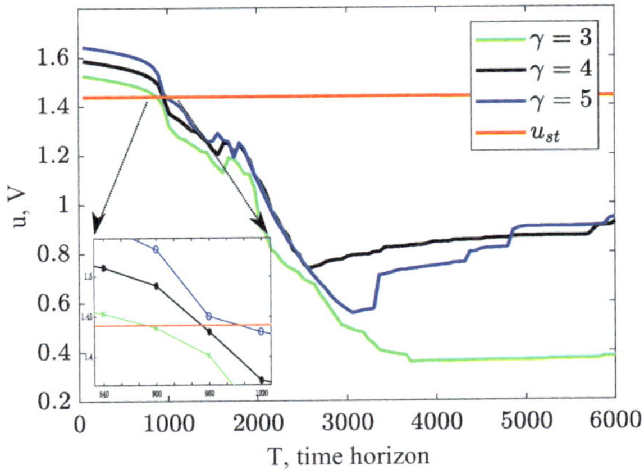

Fig. 5.6 The maximum absolute value of the control function $u(t)$: $\max_{t\in[0,T]} |u(t)|$

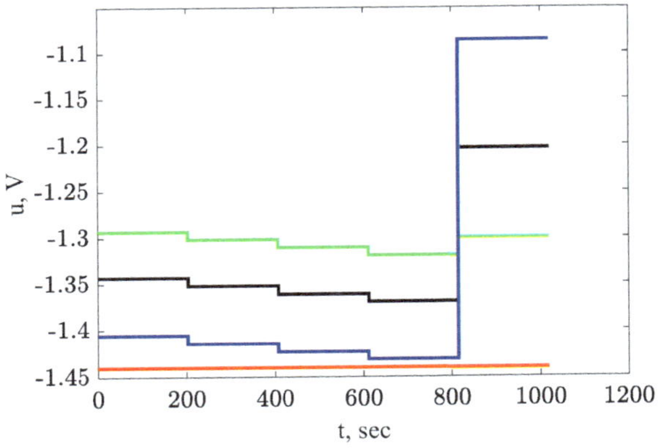

Fig. 5.7 The piecewise control function $u(t)$; $T = 1020$ sec

Figures 5.6, 5.7 illustrate the behavior of the optimal control law. Figure 5.6 shows the maximum absolute value of $u(t)$. This allows us to estimate T_{min}. Fix $\varepsilon = 10^{-2}$. Note that relaxing the temperature minimization (making γ smaller) let us achieve smaller values of T_{min}: T_{min} for $\gamma = 3$ is less then 10^3 sec, but $T_{min} > 10^3$ sec for $\gamma = 4, 5$. In Fig. 5.7, the piecewise optimal control laws for $\tilde{T} = 1020$ sec are given. The control constraint is depicted by the red line. As can be seen for the chosen time step δT_0, this value $\tilde{T}$ may be taken as T_{min} for $\gamma = 5$.

Figures 5.8, 5.9 show the terminal temperature distribution for the weighting coefficients $\gamma = 3, 4, 5$. Here, $T = 1020$ sec. Although the control laws are fairly different, see Fig. 5.7, these distributions in the controlled cylinder $\mathcal{V}_1$ are close to each other. Note that the distinction between the results of the considered control laws is more pronounced in the uncontrolled cylinder $\mathcal{V}_2$ (negative values of z), which is used as a heat sink. In Fig. 5.9, we may compare the initial steady state (dashed line) with the distributions obtained due to natural cooling (red curve) and due to the optimal control law (black curve) for $\gamma = 4$. Note the overheating of the uncontrolled

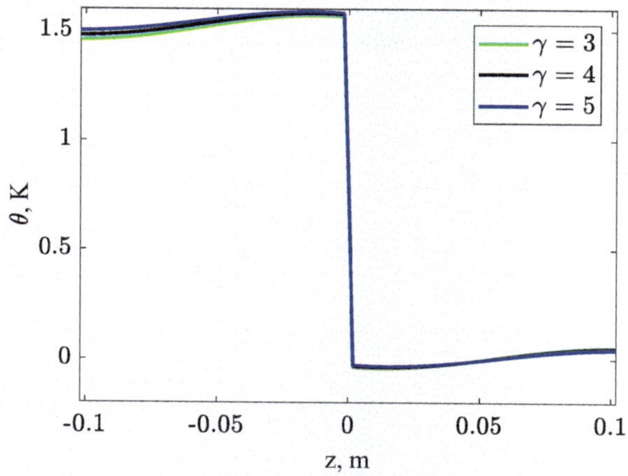

Fig. 5.8 The terminal temperature distribution

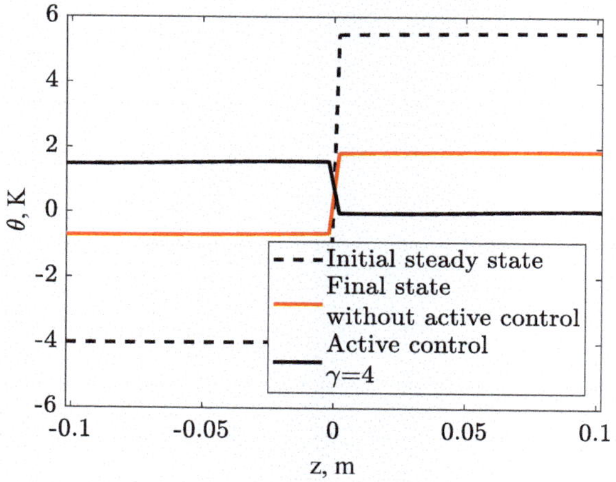

Fig. 5.9 The initial and terminal temperature distribution for the controlled and uncontrolled processes

cylinder $\mathcal{V}_2$. The time T is too short to get to the zero state in $\mathcal{V}_1$ via natural cooling and the actuator transfers the energy from $\mathcal{V}_1$ to $\mathcal{V}_2$ by heating it.

Figure 5.10 demonstrates the nonlinear essence of the studied thermoelectric system. The total input voltage involves both feedforward control law $u(t)$ and the feedback linearization signal $S[\tilde{\theta}]$ according to (5.2). Although $u(t)$ is piecewise constant, the full input voltage $U(t)$ is more sophisticated, cf. Fig. 5.7. As the electric current in the PE Ru is proportional to the voltage u, the term $S[\tilde{\theta}]$ does not influence on the Joule heat loss.

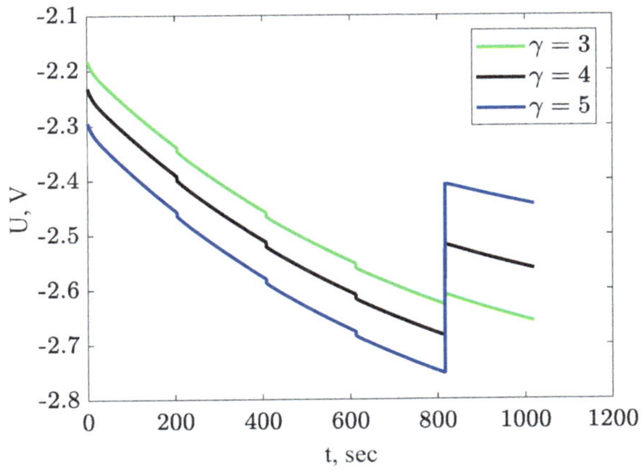

Fig. 5.10 Total input signal with feedback

5.7 Conclusions

In this paper, we study the constrained time-optimal problem of achieving a prescribed temperature distribution in a thermoelectric solid system. We consider a system consisting of two identical cylinders with a thin Peltier element between them. Although the entire system is actuated by the thermoelectric converter, we try to reach the desired state only in one of them, treating the other cylinder as a heat sink. Assuming that the optimal control law is sought in the class of piecewise constant functions, we reduce the nonlinear PDE system to its finite-dimensional approximation by utilizing eigenfunction decomposition. The constraints are implemented via a penalty term in the cost functional. Next, we vary the time horizon finding an optimal control law with applying the gradient descend method at each step of the variation. On each iteration of the gradient descend, the direct problem is solved explicitly. It has been shown that this combined numerical-analytical approach allows us to reach a rather small neighborhood of the prescribed state without violation of the constraints and estimate from above the minimal control time. We plan to verify proposed strategy and the finite-dimensional approximation exploited by implementing the designed control laws in a FEM setting. Next, we aim at controlling a structure consisting of several heat conducting bodies with Peltier elements between them.

Acknowledgements This work was supported by the Ministry of Science and Higher Education within the framework of the Russian State Assignment under contract 124012500443-0.

References

1. Tikhomirov DA, Trunov SS, Kuzmichev AV, Rastimeshin S, Shepovalova O (2020) Energy-efficient thermoelectric unit for microclimate control on cattlebreeding premises. Energy Rep 6:293–305. https://doi.org/10.1016/j.egyr.2020.08.052
2. Mironova A, Haus B, Zedler A, Mercorelli P (2020) Extended Kalman filter for temperature estimation and control of Peltier cells in a novel industrial milling process. IEEE Trans Ind Appl 56(2):1670–1678. https://doi.org/10.1109/TIA.2020.2965058
3. Fan X, Sun H, Yuan Z, Li Z, Shi R, Razmjooy N (2020) Multi-objective optimization for the proper selection of the best heat pump technology in a fuel cell-heat pump micro-CHP system. Energy Rep 6:325–335. https://doi.org/10.1016/j.egyr.2020.01.009
4. Moria H, Pourhedayat S, Dizaji HS, Abusorrah AM, Abu-Hamdeh NH, Wae-hayee M (2021) Exergoeconomic analysis of a Peltier effect air cooler using experimental data. Appl Therm Eng 186:116513. https://doi.org/10.1016/j.applthermaleng.2020.116513
5. Kwan TH, Ikeuchi D, Yao Q (2019) Application of the Peltier sub-cooled trans-critical carbon dioxide heat pump system for water heating—modelling and performance analysis. Energy Convers Manag 185:574–585. https://doi.org/10.1016/j.enconman.2019.01.104
6. Tijani IB, Al Hamadi AA, Al Naqbi KA, Almarzooqi RI, Al Rahbi NK (2018) Development of an automatic solar-powered domestic water cooling system with multi-stage Peltier devices. Renew Energy 128:416–431. https://doi.org/10.1016/j.renene.2018.05.042
7. Badescu V (2021) Self-driven reverse thermal engines under monotonous and oscillatory optimal operation. J Non-Equilib Thermodyn 46(3):291–319. https://doi.org/10.1515/jnet-2020-0103
8. Honc D, Doležel P, Merta J (2019) Thermal process control using neural model and genetic algorithm. In: Silhavy R, Silhavy P, Prokopova Z (eds) Intelligent systems applications in software engineering. CoMeSySo 2019. Advances in Intelligent Systems and Computing, vol 1046. Springer, Cham. https://doi.org/10.1007/978-3-030-30329-7_35
9. Gavrikov A, Kostin G, Aschemann H, Rauh A (2020) Modeling and control of a thermoelectric structure with a Peltier element subject to external disturbances. IFAC-PapersOnLine 53(2):7771–7776. https://doi.org/10.1016/j.ifacol.2020.12.1542
10. Gavrikov A, Kostin G, Knyazkov D, Rauh A, Aschemann H (2019) Experimental validation of a nonlinear model for controlled thermoelectric processes in cylindrical bodies. In: 24th international conference on methods and models in automation and robotics (MMAR), pp 558–563. IEEE. https://doi.org/10.1109/MMAR.2019.8864613
11. Knyazkov D, Kostin G, Gavrikov A, Aschemann H, Rauh A (2019) FEM modeling and parameter identification of thermoelectrical processes in cylindrical bodies. In: 24th International conference on methods and models in automation and robotics (MMAR), pp 501–506. IEEE. https://doi.org/10.1109/MMAR.2019.8864704
12. Kostin GV, Rauh A, Gavrikov A, Knyazkov D, Aschemann H (2019) Heat transfer in cylindrical bodies controlled by a thermoelectric converter. IFAC-PapersOnLine 52(15):139–144. https://doi.org/10.1016/j.ifacol.2019.11.664
13. Gavrikov A, Kostin G (2020) A nonlinear model of heat transfer for cylindrical bodies controlled by a thermoelectric converter. In: 15th International conference on stability and oscillations of nonlinear control systems (Pyatnitskiy's Conference) (STAB), pp 1–4. IEEE. https://doi.org/10.1109/STAB49150.2020.9140683
14. Gavrikov A, Kostin G, Knyazkov D, Rauh A, Aschemann H (2019) Parameter optimization of control with feedback linearization for a model of thermoelectric processes in cylindrical bodies. In: 24th International conference on methods and models in automation and robotics (MMAR), pp 342–347. IEEE. https://doi.org/10.1109/MMAR.2019.8864725
15. Gavrikov A, Kostin G (2023) Feedforward optimal control with constraints for a cylindrical thermoelectric system actuated by a Peltier element. In: Indeitsev DA, Krivtsov AM (eds) Advanced problem in mechanics III. APM 2021. Lecture Notes in Mechanical Engineering Cham: Springer, pp 95–107. Springer. https://doi.org/10.1007/978-3-031-37246-9_8

Chapter 6
Nonlinear Dynamics of the Primary Oscillation Circuit of a MEMS Gyroscope Under the Action of Phase-Locked Loop and Automatic Gain Control Systems

Dmitry Indeitsev, Vasilisa Igumnova, Alexey Lukin, and Ivan Popov

Abstract In this paper, we study a phase locked loop system operating in conjunction with an automatic gain control system to control the primary oscillations of high-quality micromechanical resonators using an RR-type gyroscope as an example. In order to measure the angular velocity of an object with high accuracy, it is necessary to ensure the stability of the amplitude and frequency of the primary oscillations of the sensitive element. Automatic gain control (AGC) is necessary to maintain the amplitude of the primary oscillations at the initially set level.The task of the phase-locked loop (PLL) system is to maintain the frequency of the voltage-controlled oscillator at a value equal to the resonant frequency of the primary oscillations. A complete system of equations of motion of the sensing element was written, taking into account the control loops of the PLL and AGC. Separately, the axis of primary oscillations in control loops is considered, to which the averaging method is applied to obtain the evolution of motion in slow variables for verification of direct numerical simulation. Stability conditions for the resulting stationary regime are obtained in the linear approximation. The questions of accuracy of various methods of numerical solution of differential equations in Matlab are considered. The output characteristics of control systems are presented, as well as the dynamics of the sensitive element along the primary axis of motion. A study of the PLL system was carried out in terms of such characteristics as: speed, delay area in the operating mode, as well as resistance to changes in the parameters of the control system.

Keywords Nonlinear dynamics · MEMS-gyroscope · PLL · AGC

D. Indeitsev · V. Igumnova (✉) · A. Lukin · I. Popov
Peter the Great St.Petersburg Polytechnic University (SPbPU), St. Petersburg, Russia
e-mail: igumnovavs@mail.ru

D. Indeitsev
Institute for Problems in Mechanical Engineering of the Russian Academy of Sciences, Moscow, Russia

H. Altenbach and V. Kuzkin (eds.), *Advanced Problems in Mechanics*,
https://doi.org/10.1007/978-3-031-80103-7_6

6.1 Introduction

The principle of operation of a micromechanical vibratory gyroscope is the transfer of mechanical energy between two modes of oscillation of a sensitive element (SE), caused by the action of Coriolis inertia forces. In the simplest case, the SE consists of one moving mass—a rigid body on elastic suspensions—with two orthogonal working modes of vibration. In the angular velocity measurement mode, harmonic oscillations are excited along the primary axis by an electrostatic motor. In a linear approximation, the amplitude of the secondary oscillations caused by the Coriolis force is proportional to the magnitude of the component of the angular velocity of the moving object along the sensitivity axis of the sensor.

High requirements for the stability of the operation of resonant sensors (in particular, sensitive elements of microelectromechanical vibrating gyroscopes) lead to the need to use feedback control systems that ensure the constancy of the resonator oscillation amplitude and its frequency tuning to resonance. The first of these tasks, as a rule, is solved using an automatic gain control system [1–4]. To solve the second problem, phase-locked loop systems are widely used [5–7].

Mathematical modeling of the dynamics of MEMS resonators under the conditions of operation of such control systems is of considerable theoretical and applied interest. A large number of scientific papers have been devoted to various aspects of this problem. In [8], an analytical study of stationary oscillatory processes for a mechanically linear resonator was carried out, taking into account the action of the AGC system. In [9] the dynamics of the resonator under the action of the PLL system has been studied.The region of stability of the required resonant mode of oscillations in the space of parameters of the control system is found. The dynamics of the resonator was calculated by a direct numerical method, taking into account the nonlinearity of the elastic restoring force. A qualitative study of the dynamics of a mechanically linear resonator under the combined action of AGC and PLL systems was carried out in [10–12]. One of the most promising and modern solutions to deal with sharp peaks in the amplitude-frequency characteristics, namely, with amplitude disruptions, is the use of parametric excitation, which is characterized by a smooth flat frequency response [13, 14].

This work is devoted to the study of the dynamics of the sensitive element of a high-quality vibrational micromechanical gyroscope in the PLL and AGC circuit, as well as to the study of the selection of control parameters and variation of parameters for the stability of the oscillatory system. The averaging method was used to construct the system in slow variables. Conditions for the stability of the found stationary regimes are obtained. Conclusions are given on the main characteristics of the system: speed, delay area in the operating mode, as well as resistance to changes in the parameters of the control system.

6.2 Formulation of the Problem

The sensitive element of the RR-type gyroscope performs rotational movements around the axes of primary oscillations (PO) and secondary oscillations (SO). The diagram of a micromechanical gyroscope (MMG) and a sensitive element is shown in the Fig. 6.1a, b.

In the simplest case, the MMVG contains one moving mass (MM), the dynamics of which makes it possible to determine the Coriolis inertia force acting on the system and, consequently, the component of the angular velocity of the moving object along the sensitivity axis. In the angular velocity measurement mode, harmonic oscillations are excited along the primary X-axis by an electrostatic motor. The amplitude of the secondary oscillations caused by the Coriolis force along the Y-axis is proportional to the magnitude of the angular velocity component along the sensitivity axis Z.

Systems for generating primary oscillations, in particular, PLL-AGC systems, can have quite different implementations [15]. The considered circuit of PLL and AGC has the form [10] (Fig. 6.2).

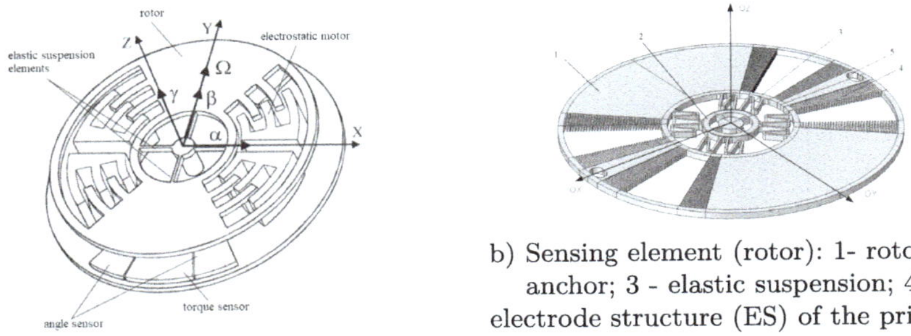

a) Scheme MMG

b) Sensing element (rotor): 1- rotor; 2 - anchor; 3 - elastic suspension; 4 – electrode structure (ES) of the primary axis; 5 - holes for stoppers

Fig. 6.1 Scheme MMG and SE

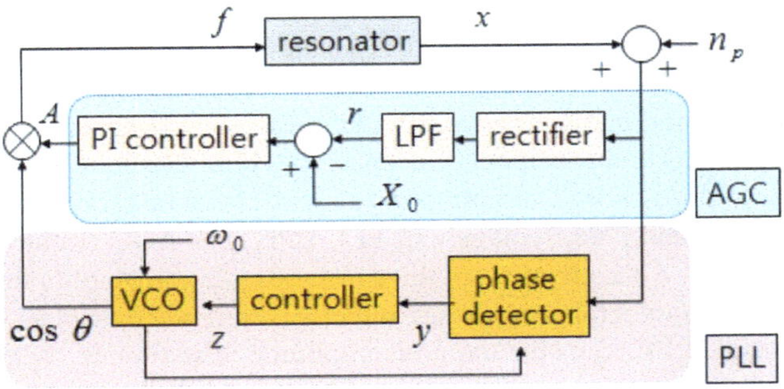

Fig. 6.2 PLL and AGC circuit

The PLL circuit contains a phase detector, a low-pass filter (LPF), a regulator, and a voltage-controlled reference oscillator (VCO) [7, 16]. The circuit is a closed control system, the task of which is to provide the required phase difference between the output signal of the capacitance-voltage converter of the resonator displacement sensor and the VCO reference signal. The amplitude channel that implements the AGC scheme consists of an amplitude detector and a PI-controller.

The dynamics of the MMG sensitive element in the PLL-AGC system is described by the following system of equations [12]:

$$
\left\{
\begin{aligned}
& J_\alpha \ddot{\alpha} + b_\alpha \dot{\alpha} + k_{\alpha 1}\alpha + k_{\alpha 3}\alpha^3 \\
& = -k_{\alpha\gamma}\alpha\gamma^2 + \left(J_\gamma - J_\beta\right)\left(-\dot{\gamma}^2 + 2\Omega_2\alpha\dot{\gamma} + \Omega_2^2\right)\alpha \\
& -\Omega_2\left(J_\gamma + J_\alpha - J_\beta\right)\left(1 - \tfrac{\gamma^2}{2}\right)\dot{\gamma} \\
& +\tfrac{R\varepsilon_A}{2}\left[\tfrac{S_1 U_1^2}{(d-R\alpha)^2} - \tfrac{S_2 U_2^2}{(d+R\alpha)^2}\right]\cdot\left(1 - \tfrac{\gamma^2}{2}\right) \\
& J_\gamma \ddot{\gamma} + b_\gamma \dot{\gamma} + k_{\gamma 1}\gamma + k_{\gamma 3}\gamma^3 \\
& = A\cos\theta - k_{\alpha\gamma}\gamma\alpha^2 + \left(J_\alpha - J_\beta\right)\Omega_2^2\gamma \\
& +\left(J_\alpha - J_\beta + J_\gamma\right)\left(\Omega_2\dot{\alpha} + \alpha\dot{\alpha}\dot{\gamma} - \tfrac{\Omega_2\gamma^2\dot{\alpha}}{2}\right) \\
& +\left(J_\alpha - J_\beta - J_\gamma\right)\Omega_2\alpha\gamma\dot{\gamma} \\
& +\tfrac{R\varepsilon_A}{2}\left[\tfrac{S_1 U_1^2}{(d-R\alpha)^2} - \tfrac{S_2 U_2^2}{(d+R\alpha)^2}\right]\alpha\gamma \\
& \dot{\theta} = \omega_0 + K_{VCO} z \\
& \dot{z} = K_I^{PLL} y \\
& \dot{y} = \lambda^{PLL}\left(K_G \gamma\cos\theta - y\right) \\
& A = K_p\left(X_0 - r\right) + B \\
& \dot{B} = K_I^{AGC}\left(X_0 - r\right) \\
& \dot{r} = \lambda^{AGC}\left(\tfrac{\pi}{2}|\gamma| - r\right).
\end{aligned}
\right.
\tag{6.1}
$$

Oscillatory system parameters: α is amplitude of secondary oscillations (SO), γ is amplitude of primary oscillations (PO), $b_{\alpha,\gamma}$ -damping coefficients of the system along the axes SO and PO, respectively, k_{α,γ_1}, k_{α,γ_3}, $k_{\alpha\gamma,\gamma\alpha}$ are linear, cubic and cross stiffness components of an elastic suspension, ε_A is absolute permittivity of the medium in the gap between the rotor and the base, R is average electrode radius, d is gap between the electrodes on the rotor and the cover, $S_{1,2}$ is overlapping area of the MM electrode and cover, $U_{1,2}$ is control voltages on the electrodes angle sensor and torque sensor.

PLL and AGC loop parameters: θ is VCO phase ($\omega = \dot{\theta}$ is instantaneous VCO frequency), z is VCO control signal, y is phase detector output, ω_0 is natural (free) VCO frequency, K_{VCO} is VCO proportional gain, K_I^{PLL} is integrated regulator gain, K_G, λ^{PLL} are gain and pole frequency of PLL LPF, A is AGC circuit output, B is AGC control signal, r is estimation of the current oscillation amplitude, K_p, K_I^{AGC} are proportional and integral coefficients of the AGC circuit controller, λ^{AGC} is pole frequency of AGC LPF, X_0 is the required amplitude of stationary oscillations.

6.3 Mathematical Model of the System in PLL and AGC Circuits

Let us consider separately the axis of primary oscillations of the gyroscope, both in linear and non-linear formulations. The equation of motion of primary oscillations has the form:

$$J_\gamma \ddot{\gamma} + b_\gamma \dot{\gamma} + k_{\gamma 1}\gamma + k_{\gamma 3}\gamma^3 = \widetilde{A}cos\theta. \tag{6.2}$$

Let's introduce: $c_d = \frac{b_\gamma}{J_\gamma}$, $\omega_\gamma^2 = \frac{k_{\gamma 1}}{J_\gamma}$, $\beta = \frac{k_{\gamma 3}}{J_\gamma}$, A—the output signal of the AGC circuit, taking into account the non-dimensionalization. Then the equation of motion will look like:

$$\ddot{\gamma} + c_d\dot{\gamma} + \omega_\gamma^2\gamma + \beta\gamma^3 = Acos\theta. \tag{6.3}$$

The parameters of the mechanical system are given in the Table 6.1.

The mathematical model of the PO axis in the PLL-AGC system is described by the following system of equations:

$$\begin{cases} \ddot{\gamma} + c_d\dot{\gamma} + \omega_\gamma^2\gamma + \beta\gamma^3 = Acos\theta \\ \dot{\theta} = \omega_0 + K_{VCO}z \\ \dot{z} = K_I^{PLL}y \\ \dot{y} = \lambda^{PLL}\left(K_G\gamma\cos\theta - y\right) \\ A = K_p\left(X_0 - r\right) + B \\ \dot{B} = K_I^{AGC}\left(X_0 - r\right) \\ \dot{r} = \lambda^{AGC}\left(\frac{\pi}{2}|\gamma| - r\right). \end{cases} \tag{6.4}$$

For a qualitative (parametric) study of the dynamics of the system, we use the averaging method. We represent the desired solution for a mechanical link in the form:

$$\gamma = a(t)\cos\left[\theta(t) + \phi(t)\right], \tag{6.5}$$

Table 6.1 Mechanical system parameters

Parameter	Value	Units
Moment of inertia about the axis Oz, J_γ	6.9573e-13	$kg * m^2$
Natural frequency of rotation around the axis Oz, f_γ	8093.7772	Hz
Quality factor around the axis Oz, Q_γ	60000	$dimensionless$
Damping factor, b_γ	4.4226e-13	$kg * m^2/s$
Linear stiffness component, $k_{\gamma 1}$	0.0018	$kg * m^2/s^2$
Cubic stiffness component, $k_{\gamma 3}$	0.0044	$kg * m^2/s^2$

where $a(t)$—amplitude, $\theta(t) + \phi(t)$—instantaneous phase of oscillation. The instantaneous phase of the VCO is θ, thus ϕ—phase difference between resonator and VCO.

Differentiating (6.5), we get

$$\dot{y} = -a\dot{\theta} \sin(\theta + \phi) + \dot{a} \cos(\theta + \phi) - a\dot{\phi} \sin(\theta + \phi). \tag{6.6}$$

Let's demand

$$\dot{a} \cos(\theta + \phi) - a\dot{\phi} \sin(\theta + \phi) = 0, \tag{6.7}$$

then

$$\ddot{y} = -\dot{a}\dot{\theta} \sin(\theta + \phi) - a\ddot{\theta} \sin(\theta + \phi) - a\dot{\theta} \left(\dot{\theta} + \dot{\phi}\right) \cos(\theta + \phi). \tag{6.8}$$

Note that the VCO phase dynamics is described by a simple equation:

$$\ddot{\theta} = K_{VCO} K_I^{PLL} y. \tag{6.9}$$

Variable θ is "fast" and can be thought of as normalized time. Following the procedure of the averaging method, we substitute (6.5), (6.6) and (6.8) in (6.4), express from the first equation (6.4) and (6.7) $\dot{a}$, $\dot{\phi}$ and we average the right parts of the system over the explicitly incoming "time" θ. Thus, we arrive at an autonomous system of differential equations describing the evolution of slow variables $\tilde{a}$, $\tilde{\phi}$, $\tilde{z}$, $\tilde{y}$, $\tilde{B}$, $\tilde{r}$:

$$\begin{cases} \dot{\tilde{a}} = -\dfrac{\tilde{B}\sin\tilde{\phi} + K_p(X_0 - \tilde{r})\sin\tilde{\phi} + c(\omega_0 + K_{VCO}\tilde{z})\tilde{a} + K_{VCO}K_I^{PLL}\tilde{y}\tilde{a}}{2(\omega_0 + K_{VCO}\tilde{z})} \\ \dot{\tilde{\phi}} = -\dfrac{\left[(\omega_0 + K_{VCO}\tilde{z})^2 - \omega_n^2\right]\tilde{a} + \left[\tilde{B} + K_p(X_0 - \tilde{r})\right]\cos\tilde{\phi} - \frac{3}{4}\beta\tilde{a}^3}{2\tilde{a}(\omega_0 + K_{VCO}\tilde{z})} \\ \dot{\tilde{z}} = K_I^{PLL}\tilde{y} \\ \dot{\tilde{y}} = -\lambda^{PLL}\left(\tilde{y} - \frac{K_G}{2}\tilde{a}\cos\tilde{\phi}\right) \\ \dot{\tilde{B}} = K_I^{AGC}(X_0 - \tilde{r}) \\ \dot{\tilde{r}} = \lambda^{AGC}(\tilde{a} - \tilde{r}). \end{cases} \tag{6.10}$$

Possible stationary periodic regimes of the (6.4) system correspond to fixed points (equilibrium positions) of the dynamical system (6.10):

$$\tilde{y}_0 = 0, \quad \tilde{r}_0 = X_0, \quad \tilde{a}_0 = X_0, \quad \tilde{\phi}_0 = \pm\frac{\pi}{2}. \tag{6.11}$$

Stationary values $\tilde{z}$ are determined from the equation

$$(\omega_0 + K_{VCO}\tilde{z}_0)^2 - \omega_n^2 - \frac{3}{4}\beta X_0^2 = 0. \tag{6.12}$$

Finally,

$$\tilde{B}_0 = cX_0 \left[\omega_0 + K_{VCO}\tilde{z}_0\right]. \tag{6.13}$$

Of main interest is the solution for which $\tilde{\phi}_0 = -\frac{\pi}{2}$ and $\tilde{z}_0$ has a sign $sgn\,(\omega_n - \omega_0)$.

The analysis of the stability of the considered stationary mode is performed by calculating the eigenvalues of the Jacobian of the right side of the system (6.10) at the given fixed point. For a linear resonator ($\beta = 0$), the stability region in the parameter space is characterized by the following conditions:

$$\begin{cases} K_I^{AGC} < K_{I,MAX}^{AGC} = K_C^{AGC} \cdot \omega_n \left(c + \frac{K_p}{\omega_n}\right) \cdot \left(\frac{c}{2} + \lambda^{AGC}\right) \\ K_I^{PLL} < K_{I,MAX}^{PLL} = K_C^{PLL} \cdot \frac{c(\lambda^{PLL} + \frac{c}{2})}{K_G K_{VCO} X_0}. \end{cases} \tag{6.14}$$

6.4 Linear Model

In the Table 6.2 we present the parameters of the linear system, taking into account that the poles of the low-pass filter are equal to each other $\lambda_{AGC} = \lambda_{PLL}$.

Below are the results of simulating system dynamics with two different numerical solvers in Matlab. Note that the numerical method of a higher order gives a high accuracy of calculations. For the given system parameters, the Fig. 6.3 shows the results of direct numerical simulation of the equation (6.4) and the analytical solution in slow variables (6.10) by the 4th-5th order numerical method (ode45).

Figure 6.4 shows the output parameters of the PLL and AGC.

The Fig. 6.5 shows the results of direct numerical simulation of the equation (6.4) and the analytical solution in slow variables (6.10) using the 8th-9th order numerical method (ode89).

Figure 6.6 shows the output parameters of the PLL and AGC.

The region of dragging into the operating mode for the linear model from the reference value $f_0 = 8092Hz$ is in the interval $[-17Hz; +13Hz]$. Simulations were also carried out when the amplitude of primary oscillations reached 3° and 5°.

Table 6.2 Control loop parameters

$Q_z = 60000$	$\lambda_{PLL} = f_Y/10$
$f_0 = 8092$	$K_C^{AGC} = 0.02$
$K_G = [1; 1e6]$	$K_p = 1e6$
$K_{VCO} = [4e3; 1e4]$	$K_C^{PLL} = 0.2$

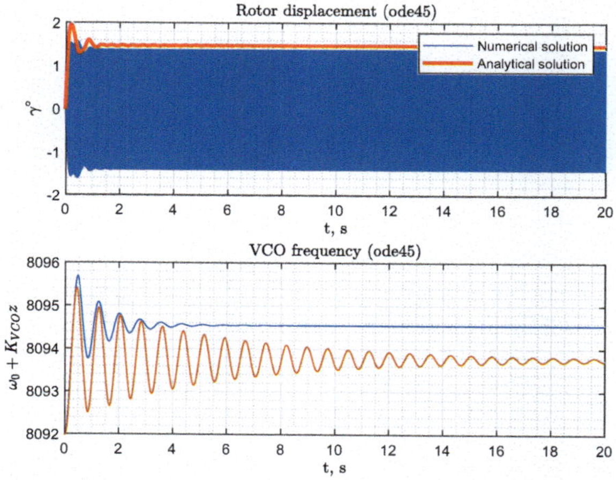

Fig. 6.3 Primary amplitude and VCO frequency

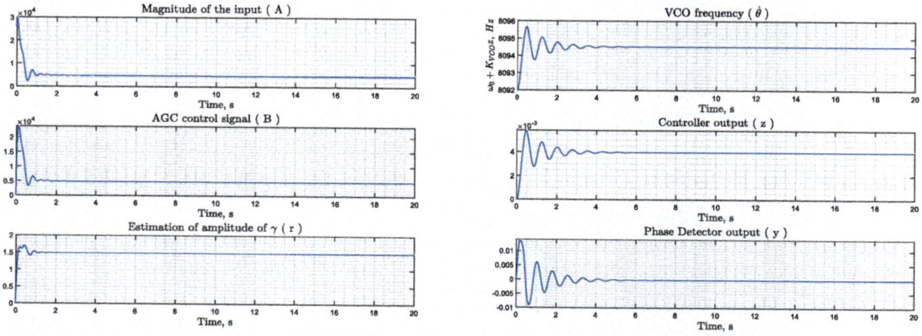

a) AGC loop parameters b) PLL loop parameters

Fig. 6.4 PLL and AGC loop parameters

Figures 6.7 and 6.8 show amplitudes of 3 and 5 degrees, as well as the VCO frequency.

An increase in the amplitude of the primary oscillations leads to large amplitudes of the secondary oscillations, which contain information about the measured angular velocity.

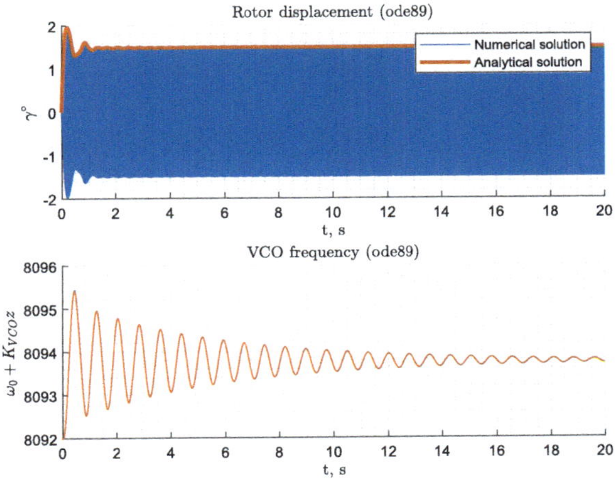

Fig. 6.5 Primary amplitude and VCO frequency

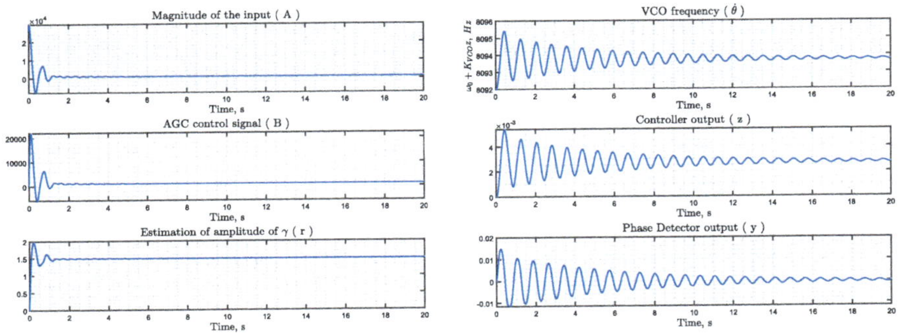

a) AGC loop parameters   b) PLL loop parameters

Fig. 6.6 PLL and AGC loop parameters

6.4.1 Mechanically Nonlinear Model

The next step in the study of SE dynamics is to take into account the nonlinearity of the mechanical link. In the Table 6.3 we present the parameters of the nonlinear system, taking into account that the poles of the low-pass filter are equal to each other $\lambda_{AGC} = \lambda_{PLL}$.

For the given system parameters, the Fig. 6.9 shows the results of direct numerical simulation of the (6.4) equation and the analytical solution in slow variables (6.10) by the 8th-9th order numerical method (ode89).

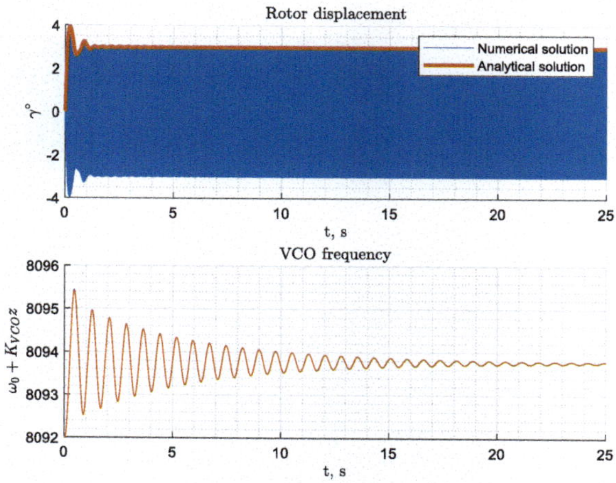

Fig. 6.7 Primary oscillation amplitude output by 3° and VCO frequency

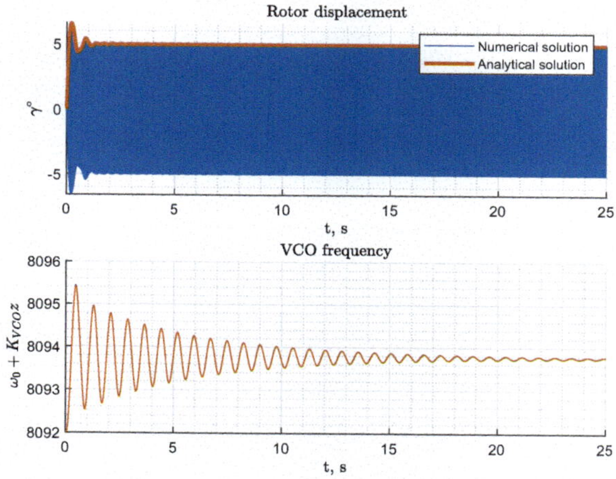

Fig. 6.8 Primary oscillation amplitude output by 5° and VCO frequency

Table 6.3 Control loop parameters

$Q_z = 60000$	$\lambda_{PLL} = f_\gamma/10$
$f_0 = 8092$	$K_C^{AGC} = 0.02$
$K_G = [1; 1e6]$	$K_p = 1e5$
$K_{VCO} = [4e3; 1e4]$	$K_C^{PLL} = 0.9$

The area of dragging into the operating mode for the nonlinear model from the reference value $f_0 = 8093Hz$ is in the interval $[-7Hz; +5Hz]$. Simulations were also carried out when the amplitude of primary oscillations reached 3° and 5°.

Figure 6.10 and shows the amplitude output at 3°, as well as the frequency of the VCO.

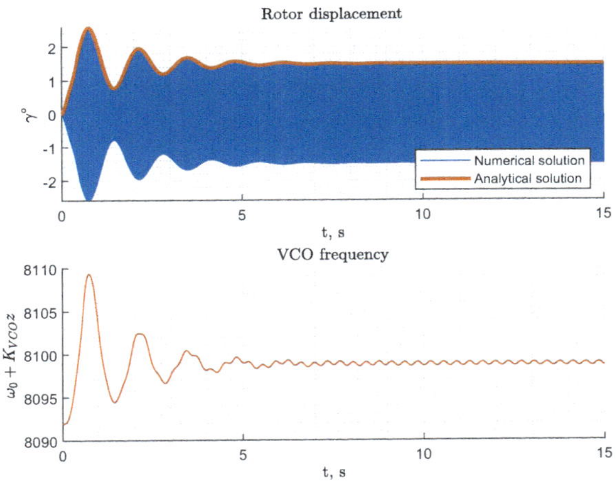

Fig. 6.9 Primary amplitude and VCO frequency

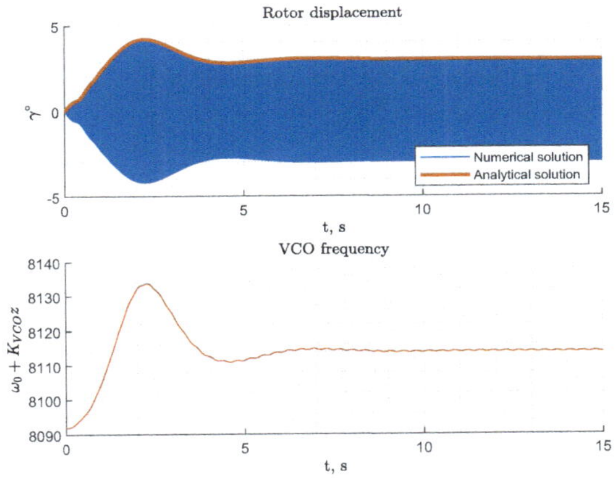

Fig. 6.10 Primary oscillation amplitude output by 3° and VCO frequency

A mechanically nonlinear system is much more complex in terms of vibration control. As can be seen from Figs. 6.9 and 6.10, the output characteristic of the VCO frequency exhibits unwanted residual oscillations that affect the accuracy of measuring the angular velocity of an object.

The Table 6.4 gives an assessment of the efficiency of the control loops in terms of speed, accuracy of the VCO frequency output, and the pull-down area in the operating mode.

Table 6.4 Evaluation of the efficiency of control loops

Compare parameter	Linear model	Mechanically nonlinear model
Minimum VCO frequency swing	$\pm 0.04Hz$	$\pm 0.22Hz$
Output time for operating regime	$20s$	$10s$
Dragging area into operating regime from the set value, f_0	$[-17Hz; +13Hz]$	$[-7Hz; +5Hz]$

6.5 Conclusion

In the present work, we study a phase locked loop system operating in conjunction with an automatic gain control system to control the primary oscillations of high-quality micromechanical resonators using an RR-type gyroscope as an example. The dynamics of the primary oscillations of the sensitive element was considered in linear and nonlinear formulations. The issue of accuracy of various numerical methods for solving differential equations in Matlab is studied and it is shown that a higher-order method gives a complete coincidence of the analytical and numerical simulation results. It is shown that for the linear model, the minimum VCO frequency swing is $\pm 0.04Hz$, the time to enter the operating mode is 20s, and the region of tightening from the initial value has the range $[-17Hz; +13Hz]$. For the nonlinear model, the minimum VCO frequency swing is $\pm 0.22Hz$, the time to enter the operating mode is 10s, and the region of tightening from the initial value has the range $[-7Hz; +5Hz]$.

The research is funded by Russian Science Foundation grant No 21-71-10009, https://rscf.ru/en/project/21-71-10009/.

References

1. Alper S, Sahin K, Akin T (2009) An analysis to improve stability of drive-mode oscillations in capacitive vibratory MEMS gyroscopes, pp 817–820.DOI: https://doi.org/10.1109/MEMSYS.2009.4805508
2. Dong L, Avanesian D (2009) Drive-mode control for vibrational MEMS gyroscopes. In: Industrial electronics, IEEE transactions on 56, pp 956–963. DOI: https://doi.org/10.1109/TIE.2008.2010088
3. Zhu H et al (2013) Drive mode control for MEMS gyroscopes with electrostatically tunable structure, pp 273–276. ISBN: 978-1-4799-1213-1. DOI: https://doi.org/10.1109/3M-NANO.2013.6737430
4. Tamara N, Barbin ES, Baranov P (2019) Amplitude control system of drive-mode oscillations of MEMS gyroscopes. In: IOP conference series: materials science and engineering 516, p 012009. DOI: https://doi.org/10.1088/1757-899X/516/1/012009
5. Wu H et al (2011) Stability analysis of MEMS gyroscope drive loop based on CPPLL, pp 45–48. ISBN: 978-1-4577-1608-9. DOI: https://doi.org/10.1109/PrimeAsia.2011.6075067
6. Demir A, Hanay S (2019) Phase-locked loop based resonant sensors: a rigorous theory and general analysis framework for deciphering fundamental sensitivity limitations due to noise

7. Lyukshonkov RG (2016) Thermal compensation in micromechanical gyroscopes with a circuit for stabilizing the amplitude of primary oscillations. In: St. Petersburg. ITMO University (2016). Dissertation for the degree of candidate of technical sciences [in russian]
8. M'Closkey R, Vakakis A (1999) Analysis of a microsensor automatic gain control loop. vol 5. , 3307–3311. ISBN: 0-7803-4990-3. DOI: https://doi.org/10.1109/ACC.1999.782377
9. Sun X, Horowitz R, Komvopoulos K (2002) Stability and resolution analysis of a phase-locked loop natural frequency tracking system for MEMS fatigue testing. J Dyn Syst Measur Control-Trans Asme 124. DOI: https://doi.org/10.1115/1.1514658
10. Park S et al (2009) Oscillation control algorithms for resonant sensors with applications to vibratory gyroscopes. In: Sensors (Basel, Switzerland) 9, pp 5952–67. DOI: https://doi.org/10.3390/s90805952
11. Indeitsev DA et al (2021) Nonlinear dynamics of MEMS resonator in PLL-AGC self-oscillation loop. Nonlinear Dyn 104(4):3187–3204
12. Indeitsev DA et al (2021) Analysis of imperfections sensitivity and vibration immunity of MEMS vibrating wheel gyroscope. Nonlinear Dyn 105(2):1273–1296
13. Miller JML et al (2019) Phase control of self-excited parametric resonators. Phys Rev Appl 12(4):044053
14. Perl T et al (2021) Control of vibratory MEMS gyroscope with the drive mode excited through parametric resonance. J Vib Acoust 143(5)
15. Saggin F (2021) Robust control for MEMS gyroscopes. PhD thesis. Lyon
16. Shahgildyan VV, Lyakhovkin AA (1972) Phase-locked loop systems. [in russian]. M.: Communication

Chapter 7
Mathematical Model of the Tapping Mode Atomic Force Microscopy in the Study of Subsurface Structures of Filled Elastomers

Roman Izyumov and Alexander Svistkov

Abstract The purpose of the work was to substantiate the methodology for constructing the subsurface layer structure of the material under study. The method is suitable for studying soft filled materials. This method consists in constructing a three-dimensional image of a rigid filler structure hiding under the surface using a phase portrait and a relief map of the sample surface. The model considers the AFM cantilever as a spring pendulum with given parameters (stiffness and resonant frequency) and calculated ones (effective mass and internal viscosity). The parameters of the numerical model describe the real experimental setup. The behavior of the material is described by an elastic model, the dissipation is given by the forces of adhesion that draw the probe into the material after contact with its surface. For a given material and experimental parameters, the obtained phase shift map is interpreted as the result of the influence of rigid objects located at a certain depth. The values of these depths are converted into the relief of hidden subsurface structures.

7.1 Introduction

AFM is a tool for studying the physical properties and structure of objects on the nanometer scale. In the so-called semi-contact mode, the cantilever taps the sample with a sharp probe located at the end of the beam [1, 2]. This makes it possible to register the phase difference (shift) between the oscillations of the cantilever tip and its base.

The task of this work was to give a theoretical proof for the hypothesis that the phase portrait may contain information about the structure of the subsurface layer. Studies are being carried out in this direction, but they use other approaches [3–5]. The advantage of the method, that we want to use, is the speed and low impact on the sample.

R. Izyumov (✉) · A. Svistkov
Institute of Continuous Media Mechanics, Perm 614013, Russia
e-mail: izumov@icmm.ru

H. Altenbach and V. Kuzkin (eds.), *Advanced Problems in Mechanics*,
https://doi.org/10.1007/978-3-031-80103-7_7

The second task was to develop an algorithm for extracting this information. The resulting technique should allow the study of soft inhomogeneous materials, such as elastomers with an active nanofiller.

7.2 Phase Shift and Effective Stiffness

What could be the reason for the phase shift? The phase shift is a reflection of energy dissipation. The reason may be: viscous, plastic behavior of the material, surface destruction, capillary phenomena, etc.

In our case, the material under study is purely elastic. Therefore, an important dissipation factor is the change in the boundaries between the media, accompanied by changes in the surface energy. Surface effects play significant role in interaction at nanometer scale. Here are some examples of works in which such effects were studied experimentally and theoretically [6–9].

On Fig. 7.1, the areas of the surface of the probe and the sample are marked in red. Before the contact, these were the "air-probe" and "air-sample" boundaries, which had energies γ_{31} and γ_{32}, respectively. After the indentation, these boundaries will disappear, and instead of them, a "probe-sample" boundary with energy γ_{12} will appear.

If the condition $\gamma_{12} < \gamma_{31} + \gamma_{32}$ is satisfied, then it will be energetically advantageous to increase the "probe-sample" boundary, and this will continue until the forces of retraction are balanced by the forces of the elastic reaction of the sample to penetration. This is a very fast process and looks like a downward jump on the indentation curve.

There is also interaction before contact, but it has a much smaller effect. In work [10], it was shown that for a probe with a tip radius of 6 nm, the jump into the polymer material will be about 0.57 nm, although in reality it can be tens of nm.

On the backward movement of the probe, there will be no jump in the model without sticking. Therefore, there will be dissipative energy losses. They will lead to the appearance of a phase shift.

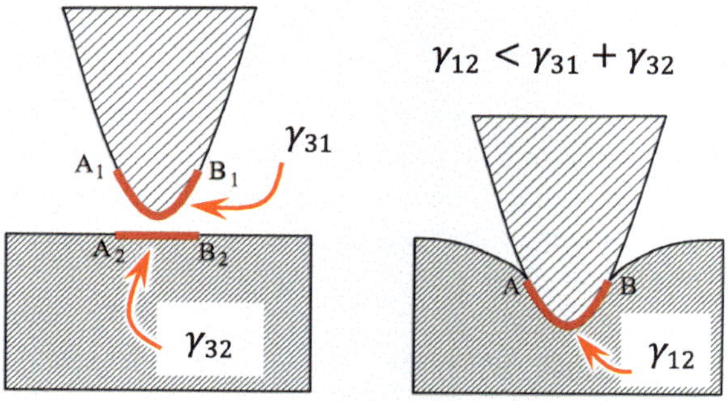

Fig. 7.1 Transformation of two interfaces (A_1B_1 and A_2B_2) into one (AB) as a result of contact

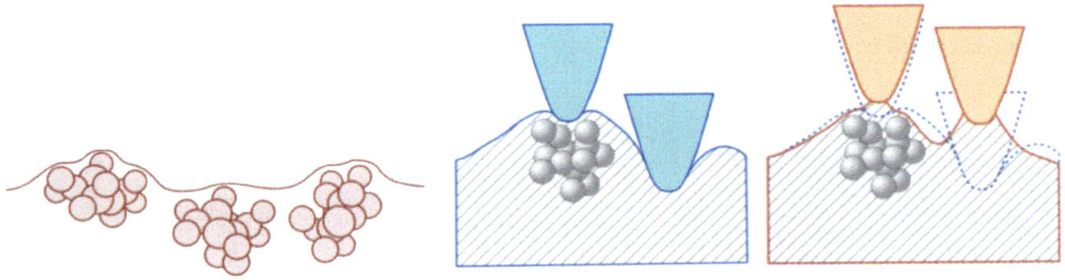

Fig. 7.2 The surface is covered with a layer of polymer material - the surface energy is the same everywhere. The factor providing a change in the phase shift will be the presence of a non-uniform distribution of "effective stiffness"

An important feature of obtaining a surface is that the experiment is carried out with a nanocomposite with an active filler. The filler particles have a high degree of interaction with the elastomer. During sample preparation, a new clean surface is created. This is achieved by cutting or tearing. The process of formation of new surfaces is associated with the growth of a crack, which is energetically favorable to bypass the aggregates of active filler particles at a certain distance. Therefore, the newly created surface will have the same surface energy everywhere (Fig. 7.2).

Therefore, the dissipation depends only on the depth of the jump made when the probe is introduced into the material. And the depth of the jump depends on the "effective stiffness of the material", that is, on how deep below the surface the filler particle is located (Fig. 7.2).

7.3 Motion Equation

The dynamic response of the cantilever-tip ensemble driven by an external and sinusoidal signal has been simulated by different authors [11–14]. The dominant contributions considered in the equation of motion of the cantilever are its elastic response, the hydrodynamic damping with the medium, the tip-sample interaction, and the excitation force.

To describe the movement of the probe, we use the model of a spring pendulum with a spring constant k, resonant angular frequency ω_0 , effective mass $m = k/\omega_0^2$.

Motion of probe z is described by nonlinear, second-order differential equation:

$$m\frac{dz^2}{d^2t} = -kd - \mu\frac{dz}{dt} - F \quad (7.1)$$

b is the position of the cantilever base,

$$b = B_0 + B\cos\omega_0 t \quad (7.2)$$

d is the cantilever deflection,

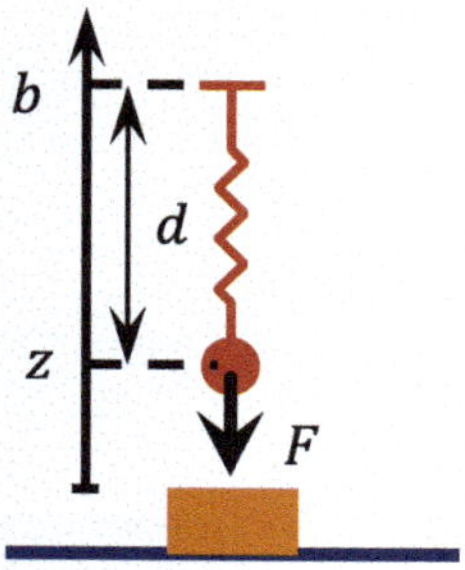

Fig. 7.3 The scheme of model—spring pendulum

$$d = z - b \tag{7.3}$$

μ is the dissipation parameter.
F is the probe-samaple interaction.

The oscillations are given to the base by eq. 7.2, which is included in eq.7.1 through relation 7.3. This is the relationship between the bend of the edge, and the positions of its tip and base (Fig. 7.3).

In the first approximation of our work, the interaction with the sample is given by the DMT model [15]:

$$F = \begin{cases} c(-z)^{1.5} - F_A & \text{if } z \le 0 \\ 0 & \text{if } z > 0 \end{cases}$$

where:

$$c = 4/3ER^{0.5}/(1 - \nu^2)$$

$$F_A = 2\pi R\gamma_{12} = 2\pi R\sqrt{\gamma_{11}\,\gamma_{22}}$$

The surface energy at the probe-sample interface is given by a simplified formula from the work of Marshall and Lee [16].

7.3.1 *Single Indentation for Samples with Different Stiffness*

Using the example of a numerical experiment that simulates indentation at a low speed, it is possible to show how the stiffness of the material affects the energy dissipation of the system. Figure 7.4 shows the indentation curves obtained on the forward and backward movements of the cantilever. Cantilever parameters is taken from [17], CSG30: $R = 10$ nm, $k = 0.6$ N/m, $f_0 = 48$ kHz.

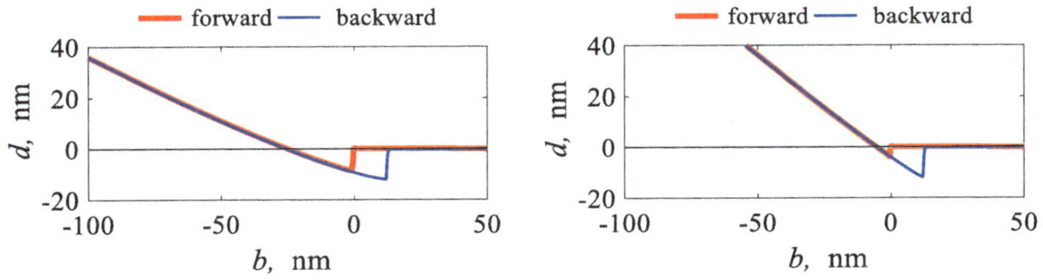

Fig. 7.4 Curves of indentation of sample with $E = 10$ MPa (left) and 100 MPa (right)

According to the results presented in Fig. 7.4, one can observe a significant effect of sample stiffness on hysteresis: when indenting a material with a modulus of $E = 10$ MPa, the hysteresis is 40% larger than for a material with $E = 100$ MPa.

7.3.2 *Tapping Mode*

To simulate the semi-contact mode, the same parameters of cantilever (CSG30) and the sample with the module $E = 10$ MPa were used.

Figure 7.5 demonstrates that we can't see clearly the influence of adhesion. This is because of the fact, that when we begin to increase a frequency of the probe oscillations in our model a very high dynamic effect of time-lag begin to dominate. At frequency of 100 Hz there is a classical picture of indentation. But probe should work at tens of thousands Hz, where we can track only phase shift (Fig. 7.6).

Figure 7.6 shows the influence of the sample stiffness on a phase shift. The additional parameters of modal are: $B = 5$ nm, $A = 44$ nm, $B_0 = A/2$.

An analysis of the model allows us to conclude that using the experimental data on the phase shift, it is possible to obtain unambiguous information about the stiffness (or effective stiffness, if we are talking about an inhomogeneous sample) of the surface layer of the sample from the obtained results of simulation with given oscillation parameters.

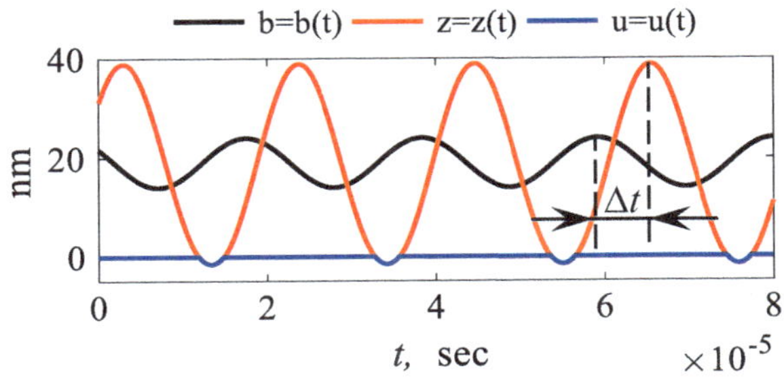

Fig. 7.5 The realization of the model in the tapping mode AFM. The black line indicates the movement of base, the red—probe position, and the blue line is a depth of indentation

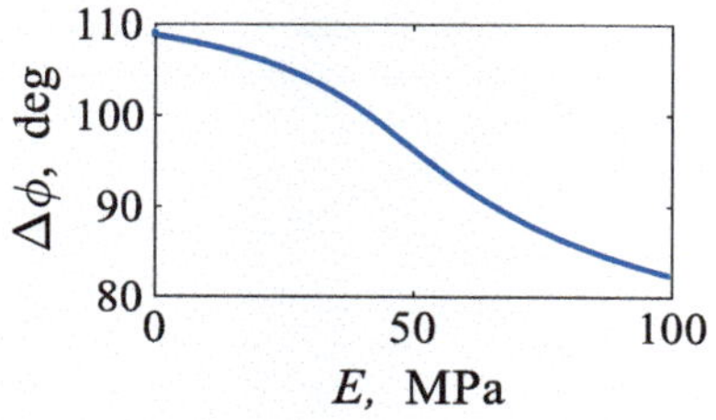

Fig. 7.6 The dependence of phase shift on the stiffness of a sample

7.4 Algorithm of Data Converting

Thus, it is possible to formulate the data transformation algorithm in the following four stages:

Stage 1. For the first stage, you need to have experimental data and get theoretical data via the proposed model (Fig. 7.7).

For clarity, let us consider the cross section of the phase shift map in the 200 nm area. Using the developed model, we calculate the dependence of the phase shift on the stiffness of the sample. The data were obtained for a cantilever with $k = 35$ N/m, $R = 10$ nm. Oscillation parameters: $B = 0.6$ nm, $A = 51.4$ nm, $B_0 = A/2$.

Stage 2. Converting is performed by replacing the phase shift values with stiffness values according to the theoretical dependence $\Delta\varphi(E)$.

$$\left.\begin{array}{c}\Delta\varphi(x)\\ \Delta\varphi(E)\end{array}\right\} \rightarrow E(x) \rightarrow E(x)/E_0 = K(x)$$

Additionally, normalizing data is produced. For this, the value of the stiffness E_0 of the unfilled sample is used. The result of the second stage is shown on Fig. 7.8.

Stage 3. In order to obtain the final result, it is necessary to conduct a study using finite element modeling, which would allow establishing a correspondence between the structure of the near-surface layer and its effective stiffness.

For research, the FE model must describe the interaction of the probe with the filled elastic material (Fig. 7.9). It is necessary to consider different sizes of inclusions

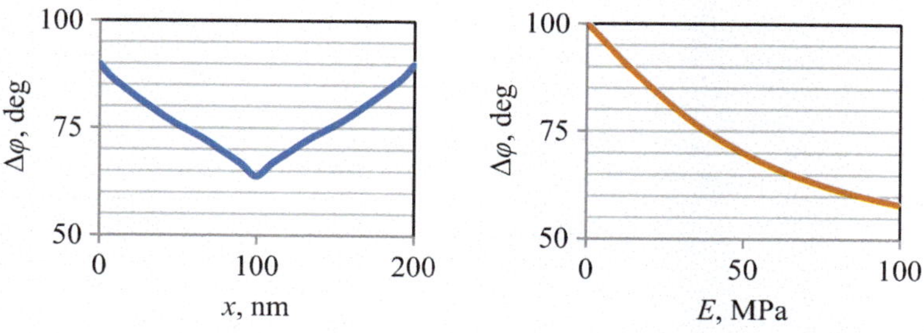

Fig. 7.7 Experimental data: phase portrait (left) and theoretical dependence $\Delta\varphi(E)$ (right)

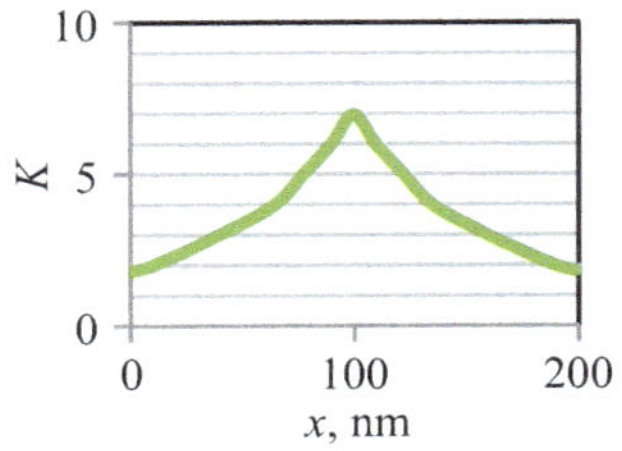

Fig. 7.8 Converted and normalized data

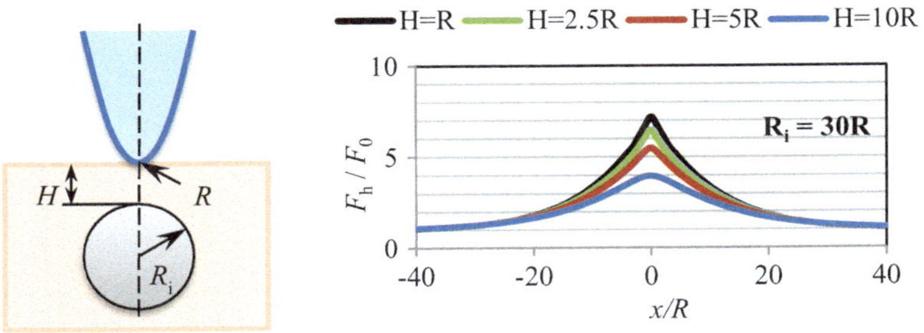

Fig. 7.9 Schematic representation of the FE model (left) and simulation result

R_i and their depth H. The probe is indented to a fixed depth h at various points of the surface x, and the force F_h required for this is determined. The result of the simulation is the force distribution over the sample surface $F_h(x)$. The final result is also normalized with respect to the value of the force F_0 for the case with an unfilled sample.

Preliminary studies [18] show that each pair of parameters (R_i, H) corresponds to a single dependence $F_h(x)/F_0$ with a unique curve shape (Fig. 7.9).

The physical meaning of the result is the effective surface stiffness $K(x) = F_h(x)/F_0$.

Stage 4. To complete, it is necessary to combine the data obtained at the 2nd and 3rd stages:

$$\left.\begin{matrix}\Delta\varphi(x)\\ \Delta\varphi(E)\end{matrix}\right\} \rightarrow E(x) \rightarrow E(x)/E_0 = K(x) = F_h(x)/F_0 \leftarrow [R_i, H]$$

$$\left.\begin{matrix}\Delta\varphi(x)\\ \Delta\varphi(E)\end{matrix}\right\} \leftrightarrow [R_i, H]$$

Thus, a correspondence was found between a pair of parameters that determine the shape and depth of the subsurface relief with the phase portrait of the surface. To construct the final relief, it is necessary to combine the obtained surface with experimental AFM data on the sample relief.

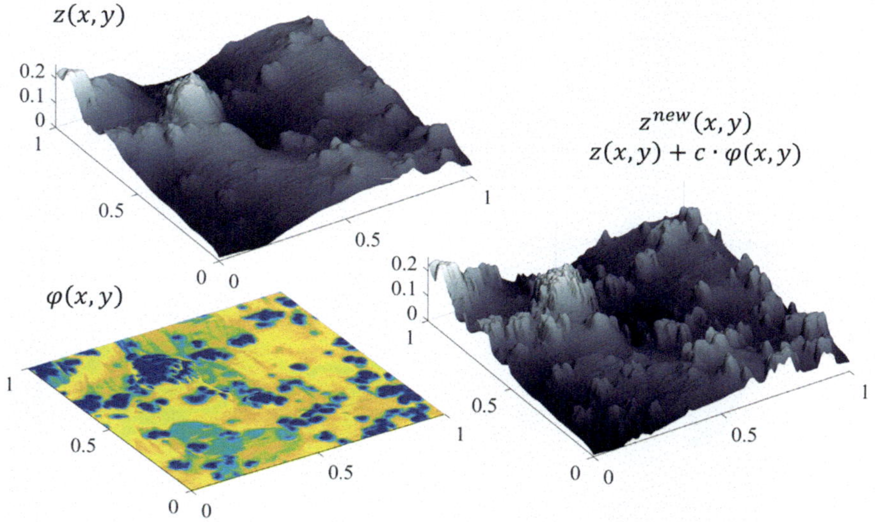

Fig. 7.10 At the top—a surface relief with a size of 1 micron $z(x,y)$, a phase portrait $\varphi(x,y)$ and a new surface $z^{new}(x,y)$ that is a linear combination of a relief and a phase portrait. In the first approximation, the phase portrait multiplied by a constant was simply added to the relief

An example of the operation of the proposed algorithm is shown in Fig. 7.10. For clarity, a simple linear transformation is used as a mathematical transformation of the phase shift.

Figure 7.10 shows how new relief details appear on relatively smooth areas. This can be compared to how the undersea relief is drawn using an echo sounder.

7.5 Conclusion

As a result of the study, a theoretical justification was given for the possibility of using a phase portrait to study the subsurface structure of soft filled materials. For this, a mathematical model of the movement and interaction of the AFM probe with the sample surface was developed. This model is also used for the proposed algorithm for converting the phase shift map and obtaining information about the structure of the subsurface layer. The scheme of operation of this algorithm is shown.

Acknowledgment The reported study was funded by RFBR and Perm Territory, project number 20-48-596013 and the Ministry of Science and Higher Education of the Russian Federation (theme No. AAAA- A20-120022590044-7).

References

1. Martin Y, Williams CC, Wickramasinghe HK (1987) Atomic force microscope–force mapping and profiling on a sub 100-Å scale. J Appl Phys 61:4723
2. Meyer G, Nabil M (1988) Erratum: novel optical approach to atomic force microscopy. Appl Phys Lett 53:1045
3. Sharahi HJ et al (2017) Contrast mechanisms on nanoscale subsurface imaging in ultrasonic AFM: scattering of ultrasonic waves and contact stiffness of the tip–sample. Nanoscale 9(6):2330–2339
4. Soliman M et al (2017) Nanoscale subsurface imaging. J Phys Condens Matter 29:173001
5. Verbiest GJ et al (2013) Subsurface-AFM: sensitivity to the heterodyne signal. Nanotechnology 24:365701
6. Svistkov AL, Izyumov RI (2020) Influence of interface phenomena on the features of interaction between the probe of atomic force microscope and soft material. Mech Mater 148:103500
7. Izyumov RI, Svistkov AL (2019) Effect of surface tension forces on changes in the surface relief of the elastomer nanocomposite. Nanosci Technol Int J 10(1):51–66
8. Rimai D, Quesnel D, Busnaina A (2000) The adhesion of dry particles in the nanometer to micrometer-size range. Colloids SurfS A PhysChemical Eng Asp 165(1–3):3–10
9. Style RW et al (2013) Surface tension and contact with soft elastic solids. Nat Commun 4:2728–2733
10. Svistkov AL, Izyumov RI (2020) Influence of intermolecular interaction force on the jump magnitude of the atomic force microscope probe during indentation of soft material. Nanosci Technol Int J 11(1):55–71
11. Eslami S, Jalili N (2012) A comprehensive modeling and vibration analysis of AFM microcantilevers subjected to nonlinear tip-sample interaction forces. Ultramicroscopy 117:31–45
12. Li Z, Shi T, Xia Q (2018) Design optimization of high performance tapping mode AFM probe. Microsyst Technol 24:979–987
13. Sahoo DR et al (2010) High-throughput intermittent-contact scanning probe microscopy. Nanotechnology 21:075701
14. Vatankhah R (2017) Nonlinear vibration of AFM microcantilevers with sidewall probe. J Braz Soc Mech Sci Eng 39:4873–4886
15. Derjaguin BV, Muller VM, Toporov YP (1975) Effect of contact deformations on the adhesion of particles. J Colloid Interface Sci 53(2):314–326
16. Marshall JS, Li S (2014) Adhesive particle flow: a discrete-element approach. Cambridge University Press, New York
17. NTMDT, CSG30, Homepage. https://ntmdt-russia.com/product/csg30
18. Izyumov RI, Sokolov AK, Svistkov AL (2021) Indentation of soft material containing filler particles. Procedia Struct Integr 32:87–92

Chapter 8
Mathematical Model of Autocrine Regulation of Osteoclasts in Bone Remodeling

A. I. Karimov, V. N. Kireev, S. I. Mustafina, O. A. Solnyshkina, A. R. Bilyalov, S. S. Chugunov, V. N. Pavlov, and I. Sh. Akhatov

Abstract Bone tissue remodeling in the human body occurs continuously in various parts of the skeleton. The process is characterized by a stage of resorption of existing mineralized tissue by osteoclasts and a stage of subsequent formation of new bone tissue by osteoblasts. The intensity of bone remodeling at each local site is regulated by numerous intracellular and extracellular parameters of cell regulation. Mathematical modeling provides an important scientific approach for integrating existing knowledge about the process of interaction of bone cells, as well as for predicting the balance between bone formation and its subsequent resorption. This study presents a mathematical model that describes changes in the populations of osteoblasts and osteoclasts, as well as subsequent changes in bone mass in the local area of bone remodeling, taking into account the diffusion movement of cells. Numerical modeling was used to study the influence of parameters of internal and external regulation of osteoclasts and osteoblasts on the process of dynamic bone formation. It is shown that the used mathematical model can describe some pathological bone diseases.

8.1 Introduction

Bone tissue remodeling is a continuous biological process occurring on an ongoing basis in various discrete areas of the skeleton. The process is characterized by the stage of resorption of existing mineralized tissue by osteoclasts and the subsequent formation of new bone tissue by osteoblasts [1]. Upon initiation of remodeling, osteoclasts differentiate from their monocytic precursors and resorb bone. Later, the

A. I. Karimov · V. N. Kireev · S. I. Mustafina (✉) · O. A. Solnyshkina
Bashkir State University, 32 Zaki Validi Str, 450076 Ufa, Russia
e-mail: sofiamustafina@mail.ru

A. I. Karimov · V. N. Kireev · S. I. Mustafina · O. A. Solnyshkina · A. R. Bilyalov · V. N. Pavlov
Bashkir State Medical University, 3 Lenina Str, 450008 Ufa, Russia

S. S. Chugunov · I. Sh. Akhatov
Skolkovo Institute of Science and Technology, 30 Bld. 1 Bolshoy Bulvd, 121205 Moscow, Russia

H. Altenbach and V. Kuzkin (eds.), *Advanced Problems in Mechanics*,
https://doi.org/10.1007/978-3-031-80103-7_8

process of differentiation of osteoblasts from mesenchymal precursors is initiated to form new bone tissue.

Bone remodeling disorders are seen in diseases such as osteoporosis, osteoarthritis, and Paget's disease [2]. In healthy people under 30 years of age, osteoblasts form the same amount of bone that was resorbed by osteoclasts. This allows you to complete the full cycle of bone formation. In people at a later age, this balance is disturbed. The regulation of bone formation [3] is complex and includes the simultaneous action of a number of factors that affect the formation or resorption of bone.

The progression of bone remodeling at each local site is regulated by numerous intracellular and extracellular factors. Transforming growth factor (TGF-β) increases bone formation through a direct effect on osteoblast synthesis [4]. The membrane protein RANKL and osteoprotegerin (OPG) are among the main regulators [5] of bone resorption. They are synthesized by osteoblasts and have an inhibitory effect on the formation of osteoclasts. Thus, cells can interact with each other using the effectors they produce, which can activate or suppress the growth of the population of osteoclasts and osteoblasts [6]. In this case, the rate of destruction is proportional to the current number of corresponding cells.

Mathematical modeling provides an important scientific approach for integrating existing knowledge about the regulation of bone cell interactions [7], as well as for predicting the balance between bone formation and its subsequent resorption [8]. In the present study, a new mathematical model that describes changes in the populations of osteoblasts and osteoclasts and subsequent quantitative changes in bone mass in the local area of bone remodeling was composed based on two well-known models described in [8, 9].

8.2 Mathematical Model

The differential equations of the mathematical model describing the dynamics of cell populations at the site of bone remodeling have the form [9]:

$$\begin{aligned} \frac{dx_1}{dt} &= a_1 x_1^{g_{11}} x_2^{g_{21}} - b_1 x_1, \\ \frac{dx_2}{dt} &= a_2 x_1^{g_{12}} x_2^{g_{22}} - b_2 x_2, \end{aligned} \tag{8.1}$$

where x_1 and x_2 are the number of osteoclasts and osteoblasts, a_1 and a_2 are the rates of osteoclasts and osteoblasts formation, b_1 and b_2 are the rates of osteoclasts and osteoblasts removal.

The four parameters g_{ij} describe the processes of producing effectors on the donor cell:

- g_{11} describes the total effect of all factors produced by osteoclasts and regulates the formation of osteoclasts,
- g_{12} describes the total effect of all factors activated by osteoclasts and regulates the formation of osteoblasts (for example, transforming growth factor TGF),
- g_{21} describes the total effect of all factors produced by osteoblasts and regulates the formation of osteoblasts (for example, insulin-like growth factors IGF),
- g_{22} describes the total effect of all factors produced by osteoblasts and regulates the formation of osteoclasts (for example, OPG and RANKL).

Populations of osteoclasts and osteoblasts under stationary conditions consist of cells that are not able to disassemble or build bone, but are able to participate in intracellular and extracellular signaling, while the rate of bone resorption and formation is proportional to the number of osteoclasts and osteoblasts, respectively. So, we obtain a differential equation describing changes in bone mass [Lemaire]:

$$\frac{dz}{dt} = -k_1 y_1 + k_2 y_2, \tag{8.2}$$

$$y_i = \begin{cases} x_i - \overline{x}_i, & ecлu \quad x_i > \overline{x}_i, \\ 0, & ecлu \quad x_i \leq \overline{x}_i. \end{cases} \tag{8.3}$$

where z is the total bone mass, k_1 and k_2 are the normalized rates of bone resorption and formation, y_1 and y_2 are the number of cells actively resorbing or forming bone, $\overline{x_i}$ is the number of cells in the stationary state.

8.3 Results of Numerical Modeling

To study the mathematical model of the interaction of osteoclasts and osteoblasts during bone remodeling, a program code was implemented in the Python programming language and software applications for a PC were implemented using the PyQt library based on the Qt framework. With the help of the created software, a dynamic process of bone formation was simulated, similar to that which occurs in Paget's disease, and experiments were carried out to determine the influence of parameters of intracellular or extracellular factors on the stability of the system. The system of differential equations was solved using the Runge–Kutta method of 4–5 orders of accuracy.

Paget's disease is a bone disease characterized by an accelerated rate of focal resorption and the formation of structural bone defects. The rate of bone remodeling in patients with Paget's disease increases with disease severity. Based on massive numerical experiments it was concluded, that the mathematical model is able to describe bone dynamics similar to pathologically bone formation in Paget's disease [10, 11] in the case when the parameters in equation [1] and [2] are chosen as follows:

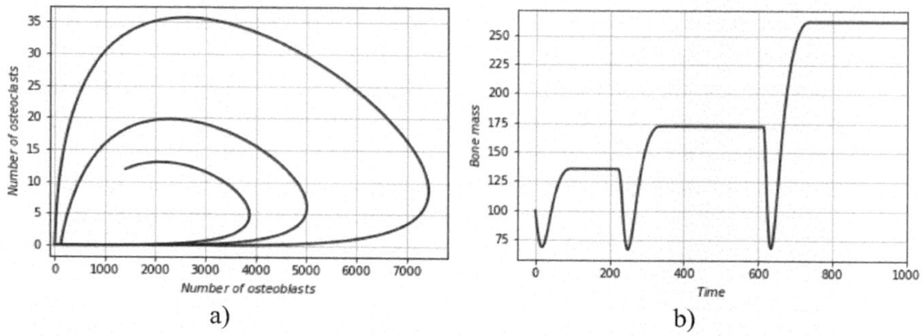

Fig. 8.1 Population dynamics of osteoclasts and osteoblasts (**a**) and changes in bone mass (**b**)

$$\alpha_1 = 7,\ \alpha_2 = 7,\ \beta_1 = 0.2,\ \beta_2 = 0.02,$$
$$g_{11} = 1.105,\ g_{12} = 1,\ g_{21} = -0.5,\ g_{22} = 0.1, \tag{8.4}$$
$$k_1 = 0.285,\ k_2 = 0.0008.$$

In the course of numerical modeling, plots of the phase portrait of the number of osteoclasts, osteoblasts and a plot of changes in bone mass were obtained.

Figure 8.1 shows that the model is in a state of oscillation, where each subsequent cycle of bone remodeling occurs with an increasing amplitude of changes in the number of osteoclasts and osteoblasts. Changes in bone mass show that with each cycle, resorption increases, followed by increased bone formation. Over time, the number of osteoblasts required to compensate for accelerated bone resorption is on the order of 5,000 cells, suggesting that the presence of preosteoblasts may limit the rate of bone formation.

To further explore the spectrum of conditions that can cause unstable oscillations, we analyzed the influence of the parameters g_{ij}. It was found that the parameters of extracellular regulation g_{21} and g_{12} have little effect on the oscillations of the system. However, an increase in the intracellular regulation of osteoclasts or osteoblasts can bring the system into the region of unstable fluctuations.

For example, when the parameter of intracellular regulation of osteoclasts g_{11} decreases to unity, the system demonstrates damped oscillations. A similar behavior of the system can be traced in the range of parameter g_{11} from 0.8 to 1.1 (Fig. 8.2).

With a slight increase in the parameter of intracellular regulation of osteoclasts up to 1.107, the system passes into the region of unstable oscillations with increasing amplitude (Fig. 8.3).

Thus, under these conditions, the intracellular regulation of osteoclasts is unique in its ability to switch the system between different modes of dynamic behavior. Intracellular factors derived from osteoclasts control the dynamic behavior of bone remodeling.

When stable fluctuations are initially observed in the system, determined by the value g_{11} (for example, equal to 1.105), an increase in the efficiency of intracellular regulation of osteoblasts g_{22} can also induce unstable fluctuations. Thus, when this

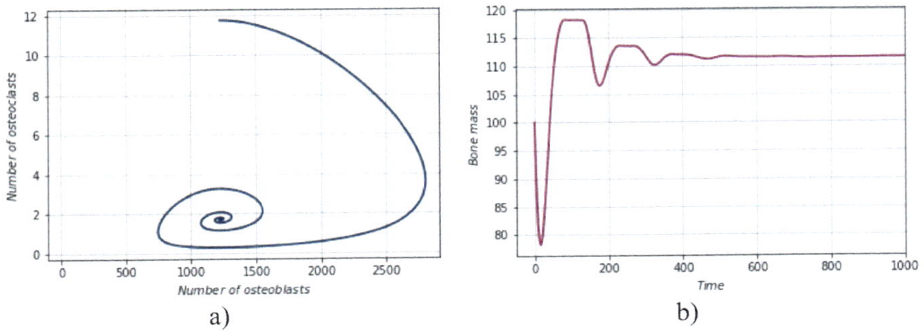

Fig. 8.2 Population dynamics of osteoclasts and osteoblasts **a** and changes in bone mass **b** at g_{11} from 0.8–1.1

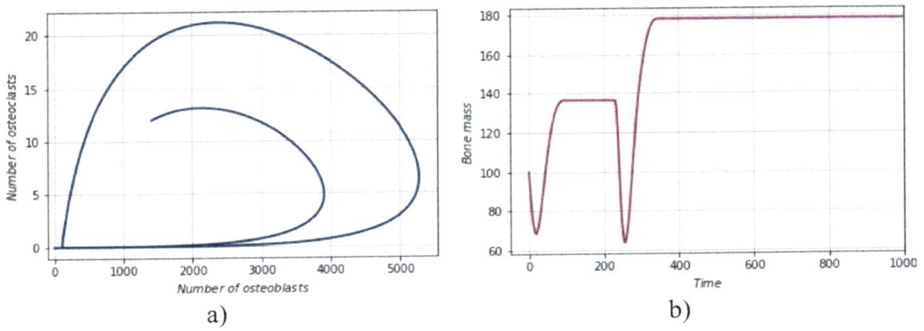

Fig. 8.3 Population dynamics of osteoclasts and osteoblasts **a** and changes in bone mass **b** with an increase in g_{11} to 1.107

parameter is increased to 0.2, the system passes into the region of unstable oscillations before reaching the stationary value of bone mass (Fig. 8.4).

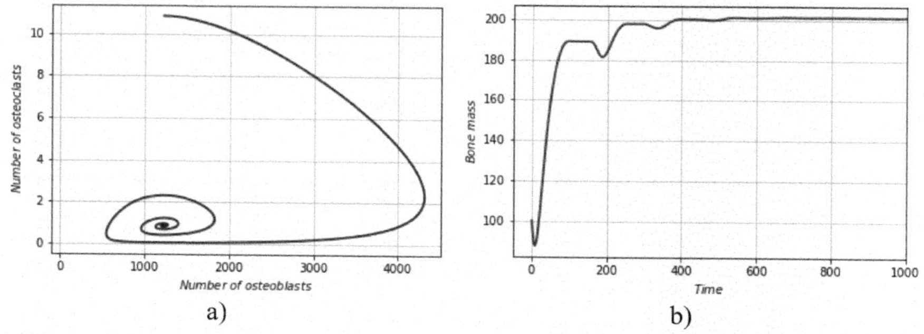

Fig. 8.4 Population dynamics of osteoclasts and osteoblasts **a** and changes in bone mass **b** with an increase in g_{22} to 0.2

8.4 Conclusion

Thus, it can be concluded that the system is most sensitive to the intracellular regulation of osteoclasts, which reflects the fact that osteoclasts disassemble the bone in the form of small groups of very active cells that are quickly recruited and then removed. However, an increase in the efficiency of intracellular regulation of osteoblasts can also induce unstable oscillations. Osteoblasts are much less active and changes in the number of osteoblasts occur more slowly; therefore, many more osteoblasts are required within a single site of bone remodeling. The model suggests that, under certain conditions, the presence of preosteoblasts may be a limiting step in the process of bone formation.

Acknowledgements The research is funded by the government of the Republic of Bashkortostan within the grant for state support of scientific research conducted under the guidance of leading scientists.

References

1. Paiva KBS, Granjeiro JM (2014) Bone tissue remodeling and development: focus on matrix metalloproteinase functions. Arch Biochem Biophys 561:74–87
2. Jones Jr RE, Austin C, Ackerman AB (1979) Extramammary paget's disease. A critical reexamination. Am J DermPathology 1(2):101–132
3. Raisz LG, Kream BE (1983) Regulation of bone formation. N Engl J Med 309(2):83–89
4. Roberts AB, Sporn MB (1991) The transforming growth factor-βs. Peptide growth factors and their receptors I. Springer, New York, NY, pp 419–472
5. Boyce BF, Xing L (2007) Biology of RANK, RANKL, and osteoprotegerin. Arthritis Res Ther 9(1):1–7
6. Merry K et al (1993) Expression of osteopontin mRNA by osteoclasts and osteoblasts in modelling adult human bone. J Cell Sci 104(4):1013–1020
7. Ayati BP et al (2010) A mathematical model of bone remodeling dynamics for normal bone cell populations and myeloma bone disease. Biol Direct 5(1):1–17

8. Lemaire V et al (2004) Modeling the interactions between osteoblast and osteoclast activities in bone remodeling. J Theor Biol 229(3):293–309
9. Komarova SV et al (2015) Mathematical model predicts a critical role for osteoclast autocrine regulation in the control of bone remodeling. Bone 33(2):206–215
10. Mills BG, Frausto A (1997) Cytokines expressed in multinucleated cells: Paget's disease and giant cell tumors versus normal bone. Calcif Tissue Int 61(1):16–21
11. Ralston SH et al (1994) Cytokine and growth factor expression in Paget's disease: analysis by reverse-transcription/polymerase chain reaction. Br J Rheumatol 33(7):620–625

Chapter 9
Optimal Control of Longitudinal Motions for an Elastic Rod with Distributed Forces

Georgy Kostin and Alexander Gavrikov

Abstract The study is devoted to mathematical modeling and optimal control design of longitudinal motions of a rectilinear elastic rod. The control inputs are a force, which is normal to the cross section and distributed piecewise constantly along the rod's axis, as well as two external lumped loads at the ends. It is assumed that the intervals of constancy in the normal force have equal length. Given initial and terminal states with a fixed time horizon, the optimal control problem is to minimize the mean mechanical energy stored in the rod. To solve the problem, two unknown functions are introduced: the dynamical potential and the longitudinal displacements. As a result, the initial-boundary value problem is reformulated in a weak form, in which constitutive relations are expressed as an integral quadratic equation. The unknown functions are both continuous in the new statement. For the uniform rod, they are found as linear combinations of traveling waves. In this case, all conditions on continuity as well as boundary, initial, and terminal constraints form a linear algebraic system with respect to the traveling waves and control functions. The minimal controllability time is found from the solvability condition for this algebraic system. After resolving the system, remaining free variables are used to optimize the cost functional. Thus, the original control problem is reduced to a one-dimensional variational problem. The Euler-Lagrange necessary condition yields a linear system of ordinary differential equations with constant coefficients supplemented by essential and natural boundary conditions. Therefore, the exact optimal control law and the corresponding dynamic and kinematic fields are found explicitly. Finally, the energy properties of the optimal solution are analyzed.

Keywords Optimal control · Dynamic systems · Model-based control · Distributed parameters

---

G. Kostin (✉) · A. Gavrikov
Ishlinsky Institute for Problems in Mechanics RAS, Moscow, Russia
e-mail: kostin@ipmnet.ru

A. Gavrikov
e-mail: gavrikov@ipmnet.ru; Alexander.Gavrikov@psu.edu
URL: https://sites.psu.edu/alexandergavrikov/

A. Gavrikov
The Pennsylvania State University, 16802 University Park, PA, USA

H. Altenbach and V. Kuzkin (eds.), *Advanced Problems in Mechanics*,
https://doi.org/10.1007/978-3-031-80103-7_9

9.1 Introduction

For many years, such classical mechanical systems as thin elastic rods or strings and related control problems have been attracting attention of scientists. This is not surprising given how many physical processes are modeled by the wave equation. A possible solution to a control problem for this model involves boundary and distributed control inputs [1]. The boundary control is often more feasible to implement in mechanical systems since its practical realization involves drivers widely used in engineering. However, it has intrinsic limitations since a finite number of inputs is used to control a continuum system. For a rod, one of the limitations is a minimal control time [2, 3].

A practical implementation of a distributed control law may suppose preliminary spatial discretization and only then be applied to a mechanical system. In our study, we assume from the beginning that the control inputs are finite-dimensional: boundary control forces are applied at the ends of the rod and a piecewise constant normal force in the cross section acts along its central line. Thus, our distributed control is discrete in space, although each of the inputs can take continuous values in time. Such control setup can be implemented, for example, with a set of piezoelectric actuators placed symmetrically along the entire length of the rod. In this work, we do not consider a detailed model of the actuators and understand forces as control inputs for simplicity. For practical implementation our control strategy may be complemented by a model of piezoelements as in [4, 5].

The assumption that the input is piecewise constant allows us to split the system into several interconnected subsystems so that each subsystem may be described by means of auxiliary traveling waves. The continuity and boundary conditions as well as control inputs interweave these functions. We use a special mesh in the time-space domain to express all the conditions in terms of a linear system, which solvability guarantees the controllability of the system. Our approach enables us to derive the exact analytical expressions for optimal control inputs, which is not typical for distributed parameter systems since usually only a numerical (approximate) solution is possible.

9.2 Statement of the Control Problem

9.2.1 Rod Kinematics and Control Forces

Let us consider longitudinal motions of a uniform rectilinear elastic rod of length $2L$ (see Fig. 9.1). The x-axis is directed along the central line with the origin at the middle of the rod. The absolute displacement of a point with the coordinate $x \in I_L$, $I_L = (-L, L)$ at the time instant $t \in I_T$, $I_T = (0, T)$ is given by a mapping $v : \Omega \to \mathbb{R}$, where $\Omega = I_T \times I_L$ is the time-space domain.

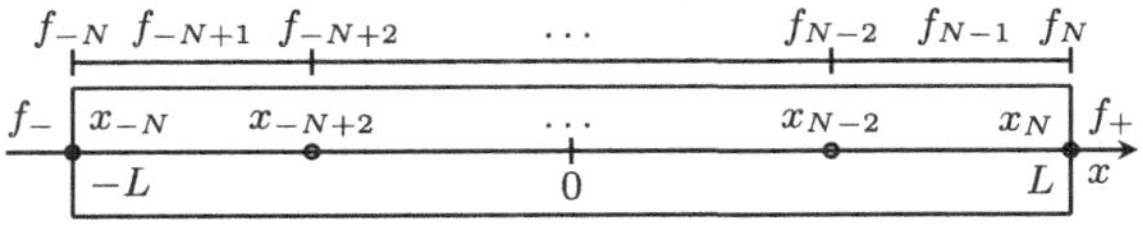

Fig. 9.1 Scheme of a rod with control elements

The rod is controlled by two external normal forces $f_\pm : I_T \to \mathbb{R}$ applied at the ends with $x = \pm L$. Additionally, the force $f : \Omega \to \mathbb{R}$ normal to the cross section tenses or compresses the elastic rod along the x-axis. We assume that the function $f(t, x)$ is piecewise constant in space. Such a load can be generated, for example, by a set of piezoelectrical actuators (control elements) attached to the rod's lateral surface [4, 5]. In our setting, these N elements have equal length and are inseparably located along the central line.

The control elements are naturally related to N space intervals I_k^x defined along the rod by points x_n given by

$$\begin{aligned} x \in I_k^x &:= (x_{k-1}, x_{k+1}), \quad k \in J_s; \\ x_n = n\lambda/2, \quad n \in J_x, \quad \lambda &= 2L/N, \quad x_{\pm N} = \pm L, \end{aligned} \tag{9.1}$$

where λ is the length of each segment. The sets of indices

$$J_s = \{1 - N, 3 - N, \dots, N - 1\}, \quad J_x = \{-N, 2 - N, \dots, N\} \tag{9.2}$$

are introduced in (9.1) to specify respectively the space intervals I_k^x and the interface points x_n.

The control function f over each segment $x \in I_k^x$ is expressed as $f(t, x) = f_k(t)$ with $k \in J_s, t \in I_T$. The renamed external loads $f_{-N-1} := f_-$ and $f_{N+1} := f_+$ complete a set of control functions f_k with the indices $k \in J_c = J_s \cup \{-N - 1, N + 1\}$. Here the supplemented index set is related to the set of control inputs (see Fig. 9.1).

9.2.2 *Initial-Boundary Value Problem*

According to the method of integro-differential relations (MIDR), [6], we define the linear momentum density $p : \Omega \to \mathbb{R}$ and the total internal normal forces in the cross section $s : \Omega \to \mathbb{R}$ as unknown variables together with the displacements v. The governing PDEs can be split into two parts. The first part is a PDE, which links the momentum and the force

$$p_t(t, x) = s_x(t, x), \quad (t, x) \in \Omega \tag{9.3}$$

according to Newton's second law. The subscripts t and x denote the partial derivatives in time and space. The second part includes two constitutive relations

$$
\begin{aligned}
&g(t,x)=0, \quad h(t,x)=0, \quad (t,x)\in\Omega, \\
&g:=\rho v_t-p, \quad h:=\kappa v_x-s+f,
\end{aligned} \tag{9.4}
$$

between the momentum p and the velocities v_t as well as between the forces s and the longitudinal strains v_x. This model contains two constants, namely, the linear density $\rho > 0$ and the tension stiffness $\kappa > 0$ of the rod.

The initial conditions are imposed on both the displacements and the momentum density by

$$
v(0,x)=v_0(x), \quad p(0,x)=p_0(x), \quad x\in I_L, \tag{9.5}
$$

where $v_0(x)$ and $p_0(x)$ are given functions. The boundary conditions are defined as follows

$$
s(t,-L)=f_{-N-1}(t), \quad s(t,L)=f_{N+1}(t), \quad t\in I_T. \tag{9.6}
$$

The continuity conditions at the inner interface points

$$
[v(t,x_n)]=0, \quad [s(t,x_n)]=0, \quad n\in J_x\setminus\{-N,N\}, \quad t\in I_T, \tag{9.7}
$$

for displacements and forces must be also taken into account. Here the square brackets denote the jump value of the variable v or s. For shortness, we define the relations (9.5)–(9.7) as interface conditions.

The jumps of derivatives $[v_x(t,x_n)]$ and the boundary values of these derivatives $v_x(t,x_{\pm N})$ depend exclusively on the differences of adjacent control functions

$$
f_n := f_{n+1} - f_{n-1}, \quad n\in J_x. \tag{9.8}
$$

The number of these functions is one less than the number of the original inputs f_k with $k\in J_c$. This means that the dynamic behavior of the system does not depend on the sum of the control signals $f_k(t)$. This value affects only the intensity of residual stresses in the rod material. To eliminate these residual stresses, let us set this sum to zero as follows

$$
\sum_{k\in J_c} f_k(t)=0, \quad t\in I_T. \tag{9.9}
$$

We define two vector spaces with elements $\boldsymbol{f}_c : I_T \to \mathbb{R}^{N+2}$ (boundary and distributed control inputs) and $\boldsymbol{f} : I_T \to \mathbb{R}^{N+1}$ (control jumps) described according to

$$
\begin{aligned}
\boldsymbol{f}_c &:= (f_{-N-1}, f_{-N+1}, \dots, f_{N-1}, f_{N+1}), \\
\boldsymbol{f} &:= (f_{-N}, f_{-N+2}, \dots, f_{N-2}, f_N).
\end{aligned} \tag{9.10}
$$

Given the vector $\boldsymbol{f}$, we can uniquely resolve the linear system of equations (9.8), (9.9) w.r.t. the entries of $\boldsymbol{f}_c$.

9.2.3 Dynamic Potential in an Equivalent IBVP Statement

Next, we define an auxiliary function or a dynamic 'potential' $r : \Omega \to \mathbb{R}$ such that

$$p(t, x) = r_x(t, x), \quad s(t, x) = r_t(t, x), \quad (t, x) \in \Omega. \tag{9.11}$$

This representation of the momentum density p and the normal forces s automatically satisfies (9.3). By excluding p and s from consideration, the state of the system is determined by two variables, kinematic v and dynamic r.

To complete properly the PDE system (9.4), we have to rewrite the interface conditions (9.5)–(9.7) in terms of the new variables v and r. As only the partial derivatives r_t and r_x appear in the governing equations (9.4)–(9.7), the variable $r(t, x)$ can be found up to an arbitrary constant $c_{0,k}$ on each subdomain $\Omega_k = I_T \times I_k^x \subset \Omega$ with $k \in J_s$.

By imposing continuity conditions $[r(0, x_n)] = 0$ with $n \in J_x \setminus \{-N, N\}$ on the initial values of dynamic variable r, the number of constants can be reduced to one (c_0) and the initial conditions (9.5) are rewritten as follows

$$v(0, x) = v_0(x), \quad r(0, x) = r_0(x) = \int_{-L}^{x} p_0(\xi)\, \mathrm{d}\xi + c_0. \tag{9.12}$$

After integrating (9.6) and (9.7) in view of (9.11), the new boundary constraints for $t \in I_T$ have the form

$$r(t, \pm L) = r_0(\pm L) + u_{\pm N \pm 1}(t). \tag{9.13}$$

The new interelement conditions are

$$[v(t, x_n)] = 0, \quad [r(t, x_n)] = 0, \quad n \in J_x \setminus \{-N, N\}. \tag{9.14}$$

The pair $u_{\pm N \pm 1}$ in (9.13) is control integrals given by

$$u_k(t) = \int_0^t f_k(\tau)\, \mathrm{d}\tau, \quad k \in J_c, \quad t \in I_T. \tag{9.15}$$

Similarly to (9.8), we can also define the jumps of control integrals $u_n(t) = u_{n+1}(t) - u_{n-1}(t)$ with $n \in J_x$. According to (9.10), (9.15), two vector-valued functions $\boldsymbol{u}_c : I_T \to \mathbb{R}^{N+2}$ and $\boldsymbol{u} : I_T \to \mathbb{R}^{N+1}$ are introduced.

9.2.4 Variational Formulation and Optimal Control Problem

Following the MIDR, an integral relation replaces constitutive PDEs to derive a variational statement in dynamics. The variational reformulation of the dynamic problem (9.4), (9.12)–(9.14) can be described as follows. Given the values of control function $\boldsymbol{u}(t)$ for $t \in I_T$, find such functions $v^*(t, x, \boldsymbol{u})$ and $r^*(t, x, \boldsymbol{u})$ in the Hilbert space $H^1(\Omega)$ that minimize the value of the constitutive functional

$$Q[v^*, r^*, \boldsymbol{u}] = \min_{v, r \in H^1(\Omega)} Q[v, r, \boldsymbol{u}] = 0, \quad Q = \int_{\Omega} q \, \mathrm{d}\Omega, \quad q = \frac{g^2}{4\rho} + \frac{h^2}{4\kappa}, \tag{9.16}$$

subject to the interface constraints (9.12)–(9.14). Here the constitutive residual function q follows from (9.4) and (9.11). We consider $\boldsymbol{u}$ as a frozen parameter in (9.16).

The following optimal control problem is considered. Find control vector-valued function $\boldsymbol{u}^*(t)$ such that the mean mechanical energy E stored in the rod over the fixed time horizon I_T reaches its minimum:

$$E[v, r, \boldsymbol{u}] \to \min_{\boldsymbol{u}, c_1}, \quad E = \frac{1}{T} \int_{\Omega} e \, \mathrm{d}\Omega, \quad e = \frac{\rho v_t^2}{4} + \frac{\kappa v_x^2}{4} + \frac{(r_t - f)^2}{4\kappa} + \frac{r_x^2}{4\rho}, \tag{9.17}$$

subject to the integral constraint (9.16), the interface conditions (9.12)–(9.14), and the continuous terminal conditions

$$v(T, x) = v_1(x), \quad r(T, x) = r_1(x) = \int_{-L}^{x} p_1(\xi) \mathrm{d}\xi + c_1, \quad x \in [-L, L]. \tag{9.18}$$

The mapping $e : \Omega \to \mathbb{R}$ denotes the linear energy density. The desired functions of displacements $v_1 : [-L, L] \to \Omega$ and momentum density $p_1 : [-L, L] \to \Omega$ completely define the terminal state of the rod, whereas the parameter c_1 does not influence on this state. The constant c_0 in (9.12) is another free parameter. A linear combination of c_0 and c_1 must be a variable parameter of optimization. Thus, without lost of generality, we can set $c_0 = 0$.

9.3 Solution of the Direct Problem

9.3.1 Representation of the Solution in D'Alembert's Form

To simplify the description of the system, we use dimensionless variables according to $v = v^*, r = \kappa \tau_* r^*, x = L x^*, t = \tau_* t^*, \tau_*^2 = L^2 \rho / \kappa$. The star superscript is further omitted. After this transformation, the length of the rod is equal to 2, whereas the length of each segment is $\lambda = 2/N$.

To analyze the optimal dynamics of a rod described in the previous section, a traveling wave representation of the problem variables (v, r) in d'Alembert's form is utilized. On each subdomain $\Omega_k = I_T \times I_k^x \subset \Omega$ with $k \in J_s$, the kinematic and dynamic variables are represented as

$$\begin{cases} v(t,x) = w_k^+(t+x) + w_k^-(t-x) \\ r(t,x) = w_k^+(t+x) - w_k^-(t-x) - u_k(t) \end{cases}, \tag{9.19}$$

where left and right traveling wave functions $w_k^\pm : I_k^\pm \to \mathbb{R}$, $w_k^\pm \in C(I_k^\pm)$, are introduced within the domains

$$I_k^\pm = \left(z_k^\pm, T - z_k^\mp\right), \quad z_k^+ = \tfrac{k-1}{2}\lambda, \quad z_k^- = -\tfrac{k+1}{2}\lambda. \tag{9.20}$$

A rather conventional coordinate representation of the traveling waves $w_k^\pm$ for $k \in J_s$ is given in the coordinate frame (z^+, z^-) rotated counter-clockwise on the angle $\frac{\pi}{4}$ w.r.t. the frame (t, x). The transformation from the old coordinates to new ones has the form $z^+ = t + x$, $z^- = t - x$. In Fig. 9.2, the $z^\pm$-axes are presented by blue lines, the orts of the new frame $\boldsymbol{i}^\pm$ with the coordinates $< \frac{1}{2}, \pm\frac{1}{2} >$ in the old system (t, x) are depicted by two blue arrows. In the new coordinates, the traveling functions $w_k^\pm$ with $k \in J_s$ depend respectively on the only argument $z^\pm$.

Although the variables (v, r) in the form (9.19) satisfy the integral constraint $Q = 0$, the solution has been given so far only for the disjoint open subdomains $\bigcup_{k \in J_s} \Omega_k$. By taking into account the initial and terminal constraints (9.12), (9.18) as well as the boundary and interelement constraints (9.13), (9.14), the solution (v, r)

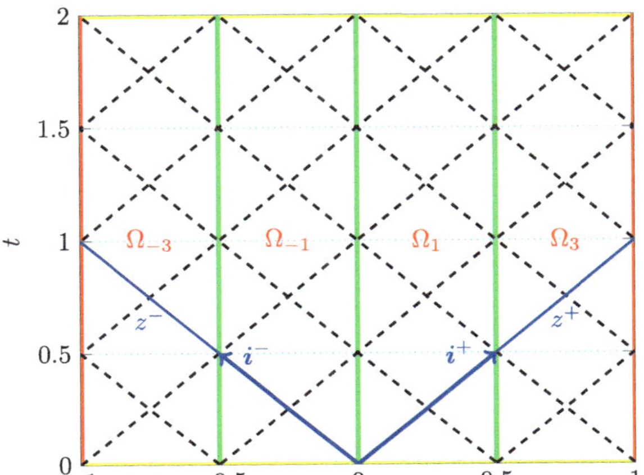

Fig. 9.2 A mesh on the dimensionless time-space domain $\Omega = (0, 2) \times (-1, 1)$ with $M = 4$, $N = 4$

needs to be extended to the set of interface edges $\overline{\Omega} \setminus \bigcup_{k\in J_s} \Omega_k$ presented in Fig. 9.2 for $N = 4$ with yellow, red, and green lines.

To resolve all the above-mentioned continuity conditions, we consider only the control time horizon T which is a multiple of the segment length λ, that is $T = M\lambda$ with $M \in \mathbb{N}$. The solution for the other values of T is found similarly but requires more complicated algorithm and is outside the scope of the paper. The mesh on the time-space rectangle $\overline{\Omega}$ is generated by the interface line segments $t = 0$, $t = T$, and $x = x_n$ for $n \in J_x$ together with the characteristic segments $z^{\pm} = \frac{i}{2}\lambda$ for $i = N + 2j$ with $j \in \mathbb{Z}$. In Fig. 9.2, the characteristic segments of the mesh with $M = N = 4$ are depicted by dashed lines. At that, $\lambda = \frac{1}{2}$ and $T = 4\lambda = 2$ for this domain.

Similarly to the space intervals I_k^x and the points x_n introduced in (9.1), the duration intervals I_l^t and the time instants t_m generated by the mesh are given by

$$\begin{aligned} t \in I_l^t &:= (t_{l-1}, t_{l+1}), \quad l \in J_d; \\ t_m &= \tfrac{1}{2}m\lambda, \quad m \in J_t, \quad \lambda = 2/N, \quad t_{2N} = T. \end{aligned} \tag{9.21}$$

The two new sets of indices are introduced in (9.21) as

$$J_d := \{1, 3, \ldots, 2M-3, 2M-1\}, \quad J_t := \{0, 2, \ldots, 2M-2, 2M\}. \tag{9.22}$$

9.3.2 *Double Indexing of D'Alembert's and Control Functions*

To extend the solution (v, r) over each closed set $\overline{\Omega}_k$, where $\Omega_k = I_T \times I_k^x$ with $k \in J_s$, we need to satisfy the continuity conditions over the initial and terminal edges

$$\begin{cases} S_k^0 = \{(t, x) \in \overline{\Omega}_k : t = 0\}, \\ S_k^T = \{(t, x) \in \overline{\Omega}_k : t = T\}, \end{cases} \quad k \in J_s, \tag{9.23}$$

and the edges parallel to the t-axis

$$S_{n,l}^x = \{(t, x) \in \overline{\Omega} : t \in I_l^t, x = x_n\}, \quad n \in J_x, \quad l \in J_d. \tag{9.24}$$

To operate with the values of $w_k^{\pm}$ on these edges in $\overline{\Omega}_k$, the open intervals $I_{k,m}^{\pm}$ of the length λ are selected on each domain $I_k^{\pm}$ from (9.20) according to

$$I_{k,m}^{\pm} = (z_{k,m}^{\pm}, z_{k,m+2}^{\pm}), \quad z_{k,m}^{\pm} = z_k^{\pm} + \tfrac{m}{2}\lambda, \ m \in J_t. \tag{9.25}$$

Here, J_t is the set of indices in (9.22), and the characteristic coordinate $z_k^{\pm}$ is given in (9.20).

The new functions $w_{k,m}^{\pm} : (0, \lambda) \to \mathbb{R}$ are defined so that

$$w^{\pm}_{k,m}(z) = w^{\pm}_{k}(z + z^{\pm}_{k,m}), \quad k \in J_s, \quad m \in J_t, \tag{9.26}$$

where the coordinate shifts $z^{\pm}_{k,m}$ are given in (9.25), the index sets J_s, J_t are introduced in (9.2), (9.22), respectively. The same procedure is applied to the control functions $u_j(t)$ defined on $t \in I_T$. Each of them is split into m edge maps $u_{j,m} : (0, \lambda) \to \mathbb{R}$, which values are equal to

$$u_{j,m}(z) = u_j\,(z + t_m)\,, \quad j \in J_x \cup J_c, \quad m \in J_t \setminus \{2M\}, \tag{9.27}$$

where the time instants t_m are given in (9.21).

After accounting for the zero initial values of u_k, the initial conditions from (9.12) in d'Alembert's form (9.19) are given on the mesh edges S^0_k introduced in (9.23) by

$$\begin{cases} w^{+}_{k,0}(z) + w^{-}_{k,0}(\lambda - z) = v_0(z^{+}_k + z) \\ w^{+}_{k,0}(z) - w^{-}_{k,0}(\lambda - z) = r_0(z^{+}_k + z) \end{cases}, \quad k \in J_s. \tag{9.28}$$

For the terminal conditions (9.18) on the edges S^T_k, we similarly arrive at

$$\begin{cases} w^{+}_{k,2M}(z) + w^{-}_{k,2M}(\lambda - z) = v_1(z^{+}_k + z) \\ w^{+}_{k,2M}(z) - w^{-}_{k,2M}(\lambda - z) = r_1(z^{+}_k + z) \end{cases}, \quad k \in J_s. \tag{9.29}$$

Note that (9.28) and (9.29) contains $4N$ relations.

The boundary conditions on $S^x_{\pm N,m}$ from (9.24) bind the traveling waves and the boundary control jumps as follows

$$\begin{aligned} w^{-}_{1-N,m+2}(z) - w^{+}_{1-N,m}(z) &= u_{-N,m}(z) + r_0(-1), \\ w^{+}_{N-1,m+2}(z) - w^{-}_{N-1,m}(z) &= u_{N,m}(z) + r_0(1) \end{aligned} \tag{9.30}$$

with $m \in J_t \setminus \{2M\}$. The number of the boundary relations is equal to $2M$. The continuity conditions on the other segments $S^x_{n,m}$ link the wave functions with the control jumps according to

$$\begin{gathered} w^{+}_{n-1,m+2} + w^{-}_{n-1,m} = w^{+}_{n+1,m} + w^{-}_{n+1,m+2}, \\ w^{+}_{n-1,m+2} - w^{-}_{n-1,m} = w^{+}_{n+1,m} - w^{-}_{n+1,m+2} + u_{n,m}, \\ n \in J_x \setminus \{-N, N\}, \quad m \in J_t \setminus \{2M\}. \end{gathered} \tag{9.31}$$

There are $2M(N-1)$ equations related to these edges.

Thus, the total number of the edge constraints is equal to $N_e = 2MN + 4N$. The equations (9.28)–(9.31) represent a linear algebraic system w.r.t. unknown wave $(w^{\pm}_{i,j})$ and control $(u_{k,l})$ functions. The total number of the unknown variables equals to $N_v = N_w + N_u = 3MN + M + 2N$ with the number of the wave functions $N_w = 2(M+1)N$ and the number of the control functions $N_u = M(N+1)$. As follows from these quantities for the system (9.28)–(9.31), the number of variables exceeds the number of equations for a potentially large value of MN. The issues of solvability of this system is discussed in the rest of this section.

9.3.3 Solvability of the System of Edge Constraints

The forced motions of an elastic rod with one interval of constancy of the distributed force, that is $N = 1$, are equivalent to motions of a rod controlled by external boundary forces $f^{\pm}$ and was described in [3]. Here, we suppose that $N > 1$ and $M > 1$. The variable surplus is equal to

$$N_s := N_v - N_e, \quad N_s(M, N) = MN + M - 2N. \tag{9.32}$$

The function N_s is positive and monotonically increasing w.r.t. M if the number of space elements N is fixed. Thus, the system (9.28)–(9.31) is underdetermined.

To prove that the underdetermined inhomogeneous linear system (9.28)–(9.31) for $M > 1$ and $N > 1$ can be resolved, we consider the following scheme. Assume that $N = 2j + 1$ with $j \in \mathbb{N}$, then:

1. the initial and terminal conditions (9.28), (9.29) are resolved w.r.t. $2N$ functions $w^{\pm}_{k,0}$ and $w^{\pm}_{k,2M}$ for $k \in J_s$;
2. the control functions $u_{\pm N,m}$ are found through the boundary conditions (9.30) for $m \in J_t \setminus \{2M\}$;
3. the interelement conditions (9.31) are resolved on the edges $S^x_{\pm n,m}$ with $m = 1, 2M - 1$ and $n > 0$, namely, the first equation in (9.31) is solved w.r.t. $w^{\mp}_{\pm n\pm 1,2}$, $w^{\pm}_{\pm n\pm 1,2M-2}$, while the second equation is solved w.r.t. the control function $u_{\pm n,l}$;
4. The equations (9.31) are resolved on the edges $S^x_{\pm n,m}$ with $m \neq 1, 2M - 1$ and $n > 0$ w.r.t. $w^{\pm}_{\pm n\pm 1,m}$, $w^{\mp}_{\pm n\pm 1,m+2}$;
5. if N is even, additionally the equations (9.31) are resolved on the central line S^x_0 w.r.t. $w^{+}_{-1,0}$, $w^{-}_{1,0}$, $w^{+}_{1,m}$ and $w^{-}_{-1,m}$ with $m \in J_t \setminus \{0, 2M\}$.

The proposed scheme for the linear equations (9.28)–(9.31) confirms their solvability for the mesh parameters $N > 1$ and $M > 1$. The unresolved variables are combined in a vector $\boldsymbol{y} : (0, \lambda) \to \mathbb{R}^{N_s}$.

The lack of variables is always present in the case $M = 1$ as follows from expression for the variable surplus N_s: $N_s = MN + M - 2N$. Indeed, equations (9.31) will include variables $w^{+}_{n-1,2}$, $w^{-}_{n-1,0}$, $w^{+}_{n+1,0}$, $w^{-}_{n+1,2}$ with $n \in J_x \setminus \{-N, N\}$, which are expressed through the initial and terminal functions v_0, r_0, v_1, r_1. It is always possible to find such a time-boundary distribution of the rod kinematic and dynamic variables v and r that some of these constraints are violated. This confirms impossibility to bring the system from an arbitrary initial state to a desired state in the critical time $T = \lambda$.

9.3.4 Continuity Conditions at Mesh Vertices

Additionally to the continuity conditions (9.28)–(9.31) on mesh edges, the corresponding linking of governing functions should be done at mesh vertices. It suffices to continuously conjugate only the free functions with their neighbors. If N is odd then the vertex conditions are given by

$$\begin{gathered} u_{n,m}(0) = u_{n,m-2}(\lambda), \quad w^{\pm}_{0,m}(0) = w^{\pm}_{0,m-2}(\lambda), \\ n \in J_x \setminus \{-N, N\}, \quad m \in J_t \setminus \{0, 2M\}. \end{gathered} \tag{9.33}$$

Thus, the total number of equations is equal to $N_b = MN + M - N + 1$. For the even number n, the following conditions are added

$$u_{0,0}(0) = 0, \quad u_{0,2M}(0) = u_{0,2M-2}(\lambda); \tag{9.34}$$

For this case, the number of equations is equal to $N_b = MN + M - N$. It should be noted that the equations of the system (9.33), (9.34) are not all necessarily linearly independent because some free functions may be coupled with dependent ones, which are, in turn, expressed through the other free variables.

9.4 Optimal Control Design

9.4.1 Decomposition of the Mean Energy

Noticing that $r_t - f = v_x$ and $r_x = v_t$ on the solution, we obtain that the energy density function e defined in (9.17) is expressed as $e = \frac{1}{2}v_t^2 + \frac{1}{2}v_x^2$ and is given on each Ω_k by

$$e = (w_k^{+\prime}(z^+))^2 + ((w_k^{-\prime}(z^-))^2 = e_k^+(z^+) + e_k^-(z^-). \tag{9.35}$$

It follows that the functional E of the rod mean mechanical energy can be split, in turn, into independent parts

$$E = \sum_{k \in J_s} \left(E_k^+ + E_k^-\right), \quad E_k^{\pm} = \frac{1}{T} \int\limits_{\Omega_k} e_k^{\pm}(z^{\pm})\, \mathrm{d}\Omega. \tag{9.36}$$

Here, the functional $E_k^{\pm}$ is defined if the wave function $w_k^{\pm}$ is given over its domain $I_k^{\pm} \in \mathbb{R}$ from (9.20), the set of segment indices J_s is introduced in (9.2). Substituting the expression for $e_k^{\pm}$ from (9.35) into $E_i^{\pm}$ we arrive at

$$E_k^{\pm} = \frac{1}{T} \int\limits_{z_k^{\pm}}^{T - z_k^{\mp}} \left(w_k^{\pm\prime}(z^{\pm})\right)^2 \Delta z_k^{\mp}(z^{\pm})\,\mathrm{d}z^{\pm}, \tag{9.37}$$

where the functions $\Delta z_i^{\pm}$, piecewise linear in $z^{\pm}$, arise as a result of primary integration over the coordinate $z^{\mp}$.

Dividing the intervals of integration $I_k^{\pm} = [z_k^{\pm}, T - z_k^{\mp}]$ in (9.35) into $M + 1$ equal parts and replacing the function $w_k^{\pm}$ for $w_{k,m}^{\pm}$ in accordance with (9.25) on the subintervals $I_{k,k}^{\pm}$ from (9.25), we arrive at

$$E = \frac{1}{T} \int\limits_{0}^{\lambda} \big(\boldsymbol{G}(z)\boldsymbol{w}'(z)\big) \cdot \big(\boldsymbol{G}(z)\boldsymbol{w}'(z)\big) dz. \tag{9.38}$$

Here, $\boldsymbol{G} : [0, \lambda] \to \mathbb{R}^{N_w \times N_w}$ is a known diagonal weighting matrix-valued function with $N_w = 2(M + 1)N$. The vector-valued function $\boldsymbol{w} : [0, \lambda] \to \mathbb{R}^{N_w}$ is introduced with its elements w_j given as follows

$$w_{2i-1} = w_{k,m}^{+}, \quad w_{2i} = w_{k,m}^{-}, \quad i = (M + 1)(k + N - 1) + m + 1. \tag{9.39}$$

9.4.2 One-Dimensional Variational Problem

In the previous section, the traveling waves $w_{k,m}^{\pm}$ have been linearly expressed through the free vector-valued function $\boldsymbol{y}$ as well as the initial and terminal states $\big(v_0(x), r_0(x)\big)$ and $\big(v_1(x), r_1(x)\big)$. It can be proven [7] that we can symbolically represent this relation according to

$$\boldsymbol{w}(\boldsymbol{y}(z), z, c_1) = \boldsymbol{A}\boldsymbol{y}(z) + c_1\boldsymbol{a} + \boldsymbol{g}(z), \quad z \in [0, \lambda]. \tag{9.40}$$

Here, $\boldsymbol{A} \in \mathbb{R}^{N_w \times N_s}$ is a known constant matrix, $\boldsymbol{a} \in \mathbb{R}^{N_w}$ is a known vector, and $\boldsymbol{g} : [0, \lambda] \to \mathbb{R}^{N_w}$ is a function defined by the initial and terminal values of v and r.

Let us also take into account the vertex conditions in (9.33), (9.34) and represent them as

$$\begin{gathered} \boldsymbol{B}_1\boldsymbol{y}(\lambda) - \boldsymbol{B}_0\boldsymbol{y}(0) = c_1\boldsymbol{b}_1 + \boldsymbol{b}_0, \\ \{\boldsymbol{B}_0, \boldsymbol{B}_1\} \subset \mathbb{R}^{N_b \times N_s}, \quad \{\boldsymbol{b}_0, \boldsymbol{b}_1\} \subset \mathbb{R}^{N_b}. \end{gathered} \tag{9.41}$$

In these terms, the optimal control problem (9.17) can be reduced to a problem of calculus of variations: Find such a function $\boldsymbol{y}^*(z)$ over $z \in [0, \lambda]$ and a constant c_1^* that minimize the functional

$$E[\boldsymbol{y}^*] = \min_{\boldsymbol{y}, c_1} E[\boldsymbol{y}] \tag{9.42}$$

Table 9.1 Dimensionless parameters

T (time)	N	M	λ (length)
2	4	4	1/2

subject to the boundary constraints (9.41).

It can be derived from (9.37) and (9.38) that that the Euler–Lagrange ODE system is given by

$$\boldsymbol{A}^{\mathrm{T}}\boldsymbol{A}\boldsymbol{y}''(z) = -\boldsymbol{A}^{\mathrm{T}}\boldsymbol{g}''(z), \quad z \in [0, \lambda]. \tag{9.43}$$

The conjugate vector-valued function to the vector $\boldsymbol{y}$ is given as

$$\boldsymbol{p} = \partial e/\partial \boldsymbol{y}' = \boldsymbol{A}^{\mathrm{T}}\boldsymbol{A}\boldsymbol{y}' + \boldsymbol{A}^{\mathrm{T}}\boldsymbol{g}'. \tag{9.44}$$

The natural boundary condition in variations is

$$\boldsymbol{p}(\lambda) \cdot \delta \boldsymbol{y}(\lambda) - \boldsymbol{p}(0) \cdot \delta \boldsymbol{y}(0) = 0. \tag{9.45}$$

Next, we take into account the variation of essential boundary constraints (9.41) according to

$$\boldsymbol{B}_1 \delta \boldsymbol{y}(\lambda) - \boldsymbol{B}_0 \delta \boldsymbol{y}(0) = \delta c_1 \boldsymbol{b}_1. \tag{9.46}$$

From (9.46), we can express the variation δc_1 and represent the natural conditions on $\boldsymbol{p}(0)$ and $\boldsymbol{p}(\lambda)$ in the form

$$\begin{gathered} \boldsymbol{p}(0) = \boldsymbol{C}_0^{\mathrm{T}}\boldsymbol{h}, \quad \boldsymbol{p}(\lambda) = \boldsymbol{C}_1^{\mathrm{T}}\boldsymbol{h}, \\ \boldsymbol{C}_i = \boldsymbol{B}_i - |\boldsymbol{b}_1|^{-2}\boldsymbol{b}\boldsymbol{b}_1^{\mathrm{T}}\boldsymbol{B}_i, \quad i = 0, 1, \quad \boldsymbol{h} \in \mathbb{R}^{N_b}. \end{gathered} \tag{9.47}$$

Finally, the system of the boundary constraints (9.41), (9.47) is solved w.r.t. $2N_s$ constants appearing after integration of (9.43) and N_b unknown components of the vector $\boldsymbol{h}$.

9.4.3 *Example of Optimal Control Design*

Let us consider an example of optimal control designed for the system with four distributed inputs ($N = 4$). The control time is equal to $T = M\lambda = 2$, what generates the mesh presented in Fig. 9.2. The mesh parameters are given in Table 9.1. The initial and terminal conditions are taken as

$$v(0, x) = \cos 3x, \quad r(0, x) = -\cos 3x, \quad v(T, x) = 0, \quad r(T, x) = c_1. \tag{9.48}$$

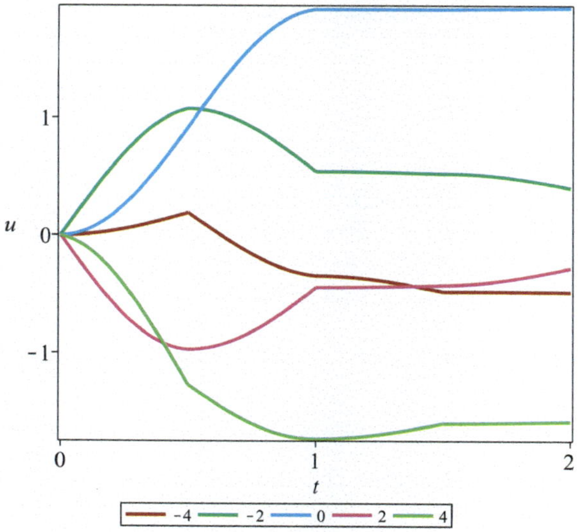

Fig. 9.3 Optimal control $u_n(t)$ for $n \in J_x$ on the dimensionless time-space domain $\Omega = (0, 2) \times (-1, 1)$ with $N = 4$

The optimal displacements $v(t, x)$ are shown in Fig. 9.4. As seen from the graph, the rod arrives to the zero state at the terminal time instant $t = 2$ both with zero energy and global displacement. The corner points of the displacement function are visible, located according to the characteristics, which are shown in in Fig. 9.2 by dashed lines.

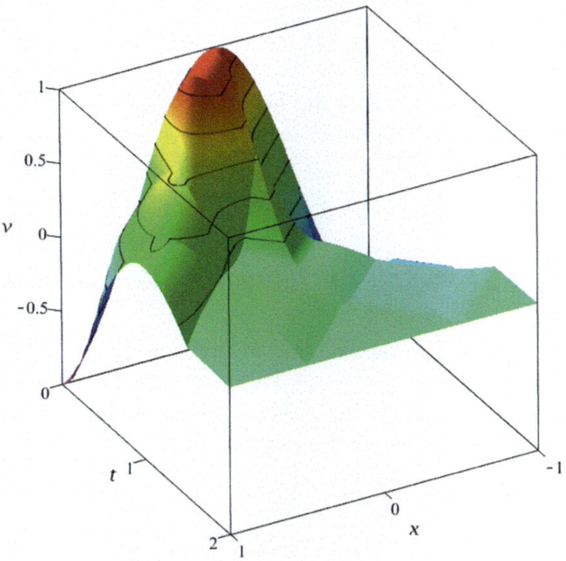

Fig. 9.4 Optimal displacements $v(t, x)$ at $N = 4$ and $T = 2$

The optimal integrals of the force jumps $u_n(t)$ with $n \in J_x = \{-4, -2, 0, 2, 4\}$, which are components of the control vector $\boldsymbol{u}(t)$, are shown in Fig. 9.3. According to the general solution of Euler–Lagrange equations and the chosen initial and terminal conditions, these integrals are linear combinations of trigonometric and polynomial functions of time. Each control map $u_n(t)$ is continuous and by definition in (9.15) has got the zero initial value $u_n(0) = 0$.

To restore the potential $r(t, x)$ and the external loads $f_{\pm}(t)$ as well as internal forces $f(t, x)$, we need first to calculate the control integrals $u_k(t)$ for $k \in J_c = \{-5, -3, -1, 1, 3, 5\}$. These functions are obtained from the optimal integrals $u_n(t)$ with $n \in J_x$. The control inputs are found as the first derivatives of control integrals $f_k(t) = u_k'(t)$ with $k \in J_c$. These functions have discontinuity at the three time points $t = \frac{1}{2}, 1, \frac{3}{2}$.

Let us analyze the change in the optimal value of the cost functional E depending on the number of elements N and control time T. Since the minimized mean energy E introduced in (9.17) is the integral of the energy density $e(t, x)$ over the domain Ω divided by the fixed parameter T, it is worthwhile to examine the dependence of the minimum of this integral $T \cdot E$ on the control parameters. In the case of the discrete time set $T = M\lambda$, we can use the mesh numbers M and N as independent parameters.

The optimal values of the energy integral $T \cdot E$ versus the mesh parameters M, N are depicted in Fig. 9.5 for the initial and terminal distribution (9.48). As seen on the graph, the functional $T \cdot E(M, N)$ is monotonically decreasing with increasing of both the number of time intervals M and the number of space segments N. After a few time steps, the rate of change of the functional w.r.t. M becomes almost negligible. In contrast, the value $T \cdot E$ continues to decline with an increase in the number of elements N.

To demonstrate the behavior of the function $T \cdot E(M, N)$ at some control time $T = \text{const}$, we may fix the ratio $\frac{M}{N}$. For example, the isochronic value $\frac{M}{N} = 1$ gives

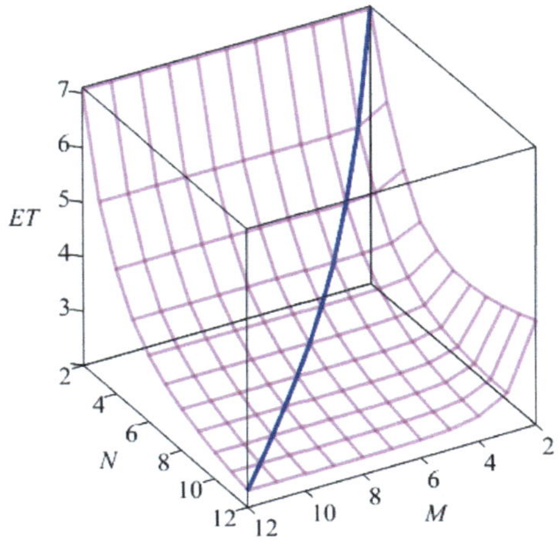

Fig. 9.5 Energy integral TE vs. the mesh parameters M, N

us the control time $T = 2$. The blue line in Fig. 9.5 presents a monotonic decline of the one-dimensional function $T \cdot E(N, N)$. The case $N = 2$ is special because the optimal value of the considered functional is the same for all $M \geq 2$. It is also worth noting the asymptotic behavior of the mapping $T \cdot E(2, N)$, which seems to tend to the limiting value. The proof of this fact will require a correct transition to the limiting case $N = \infty$ of distributed control, which is beyond the scope of this work.

9.5 Conclusions

The motion of a dynamic system under external boundary loads and internal distributed stresses has been studied. The proposed mathematical model can be related to longitudinal vibrations of a thin elastic rod controlled by forces applied along its central line together with normal forces acting at the ends. For given initial and terminal states and a fixed time horizon, the optimal control problem is to minimize the mean energy stored in the rod during the motion. In the case of equidistantly placed actuators and a uniform rod, the shortest possible time for bringing the system with any number of control elements to an arbitrary state is defined. An optimization algorithm using traveling waves is proposed to reduce the original problem to an one-dimensional variational problem with boundary conditions of a special kind. The latter is solved exactly and the optimal control inputs are found explicitly. A practical implementation of such a feedforward control usually requires additional feedback control to ensure robustness. We plan to enhance our approach by taking into account that piezoelectric elements, which can be utilized for practical realization, serve both as sensors and observers, and use that information for additional feedback stabilization.

Acknowledgements This work was supported by the Ministry of Science and Higher Education within the framework of the Russian State Assignment under contract No 124012500443-0.

References

1. Lions JL (1971) Optimal control of systems governed by partial differential equations. Springer-Verlag, New York
2. Butkovsky AG (1969) Optimal control of distributed parameter systems. Elsevier, New-York
3. Gavrikov A, Kostin G (2021) Optimal control of longitudinal motion of an elastic rod using boundary forces. J Comput Syst Sci Int 60(3):74–90. https://doi.org/10.1134/S1064230721050099
4. Kumar KR, Narayanan S (2008) Active vibration control of beams with optimal placement of piezoelectric sensor/actuator pairs. Smart Mater Struct 17(5):055008. https://doi.org/10.1088/0964-1726/17/5/055008
5. Li F, Zhang C, Liu C (2017) Active tuning of vibration and wave propagation in elastic beams with periodically placed piezoelectric actuator/sensor pairs. J Sound Vib 393:14–29. https://doi.org/10.1016/j.jsv.2017.01.038

6. Kostin GV, Saurin VV (2018) Dynamics of solid structures. Methods using integrodifferential relations. De Gruyter, Berlin
7. Kostin GV, Gavrikov AA (2022) Optimal motions of an elastic structure under finite-dimensional distributed control. J Optim Theory Appl

Chapter 10
Nonlinear Dynamics of Meshed Nanoplate Taking Into Account Self-heating

Ekaterina Krylova

Abstract In this work, a mathematical model of meshed plate nonlinear oscillations is constructed within the framework of a couple thermodynamics problem. The plate consists of an arbitrary number of densely spaced ribs families. Each family is characterized by the width of the family ribs, distance between the family ribs, and the inclination angle of the rib relative to the abscissa axis positive direction. The theory used in the work makes it possible to replace the ribs system with a continuous layer. The motion equations of plate element, boundary and initial conditions are obtained from the Hamilton's variational principle by using Kirchhoff's hypotheses. Geometric nonlinearity is taken into account according to the T von Karman's theory. Scale effects are taken into account on the basis of the micropolar theory with constrained particles rotation. The Dugamel-Nemon's hypothesis was taken into consideration. Three-dimensional hyperbolic type heat equation in a coupled formulation is considered. The heat equation is based on the Cattaneo-Vernot's hypotheses. The carbon nanoplate nonlinear dynamics under the action of a temperature field and a vibrational normal load is investigated.

Keywords Nonlinear dynamics · Carbon nanoplate · Couple thermodynamics problem · Meshed plate

10.1 Introduction

Meshed nanosized shells, cylindrical panels and plates are widely used as MEMS and NEMS elements. In particular, such structural elements as carbon nanoplates are the basis of a field-effect transistors new generation [1, 2]. Carbon-based nanostructures have a particular advantage for sensor technologies due to their potentially high sensitivity, flexibility, and reliability of physical properties.

E. Krylova (✉)
Saratov State University, 83 Astrakhanskaya Street, 410012 Saratov, Russian Federation
e-mail: kat.krylova@bk.ru
URL: https://www.sgu.ru/person/krylova-ekaterina-yurevna

H. Altenbach and V. Kuzkin (eds.), *Advanced Problems in Mechanics*,
https://doi.org/10.1007/978-3-031-80103-7_10

This necessitates the reliable mathematical models construction and the development of methods for their study to study the such structural elements behavior under the influence of various external factors. In particular, such as vibration loads and temperature effects.

The mathematical model of the meshed plate nonlinear dynamics is based on the Kirchhoff's hypotheses. But the carbon plate is a nanoscale object therefore the capabilities of the classical model will not be enough to describe the scale effects. One of the leading places among theories that allow taking into account scale effects is the micropolar (moment) theory [3, 4]. Many researchers use it to analyze the behavior of nano- and micro-sized mechanical objects [5–10]. In this article to take into account scale effects, a micropolar theory with constrained rotation of particles is considered, while it is assumed that the fields of displacements and rotations are not independent. Thus, in the model, along with force stresses, moment stresses are considered. The micro-rotations vector components are expressed in terms of the displacement vector components. The dependence of the higher-order moment symmetric tensor components on the bend-torsion tensor components is taken in the form proposed by Yang [11].

Studies of the nonlinear dynamics of NEMS and MEMS structural elements features depending on their operating conditions are very important for understanding the processes occurring in devices, for stabilizing operating modes, neutralizing harmful external influences (thermal, vibrational, etc.) [12–17]. In order to be able to analyze the nonlinear dynamics, the motion equations of the plate element are obtained taking into account the geometric nonlinearity according to the Theodore von Karman's theory.

Carbon plates, panels, nanotubes lie in the plane of scientific interests of many authors and scientific groups, however, these objects behavior is mainly considered on the basis of theories that take into account only scale effects. There are not many works devoted to the study of mesh plates, taking into account scale effects [18, 19].

In this study, object meshed structure is proposed to be taken into account on the basis of Pshenichny's theory, which assumes that densely located ribs system can be replaced by continuous layer (homogeneize), which will lead to accounting for geometry in stiffness [20].

To take into account temperature effects, the Duhamel-Neumann hypothesis is taken into account in the work. It is believed that thermal deformations are superposition of elastic deformations and thermal expansions. For a better understanding of the nanoplate behavior under temperatures action, one should consider the process of self-heating, which will lead to the need to formulate and solve the couple problem of thermodynamics. Moreover, it should be taken into account that when solving boundary value problems with a parabolic differential heat equation based on the Fourier hypothesis, the rate of heat propagation at the initial time moment tends to infinity, which in the case of bodies with small dimensions can lead to a serious error in the result.

Therefore, the mathematical model of the nanoscale meshed plate oscillations in temperature field was built taking into account the heat conduction equation based on the Cattaneo-Vernot's hypotheses [21, 22]. This made it possible to obtain a hyperbolic heat equation and remove the problem of the heat propagation infinite velocity at the initial time moment.

10.2 Problem Statement

10.2.1 Motion Equations of Continuous Shell Element

Let us assume that the isotropic homogeneous meshed plate occupies a region $\Omega = \{0 \leq x \leq c; \quad 0 \leq y \leq b; \quad -\frac{h}{2} \leq z \leq \frac{h}{2}.\}$ in the space $\mathbf{R}^3$ (Fig. 10.1). Displacement vector components, taking into account the Kirchhoff's hypotheses take the form:

$$u_x = u(x, y, t) - z\frac{\partial w(x, y, t)}{\partial x}, \quad u_y = v(x, y, t) - z\frac{\partial w(x, y, t)}{\partial y}, \quad u_z = w(x, y, t), \tag{10.1}$$

here u, v are axial displacements of the plate middle surface in the directions x, y, w is deflection. Non-zero components of the strain tensor, taking into account Karman's geometric nonlinearity take the form:

$$\begin{aligned} e_{xx} &= \frac{\partial u}{\partial x} + \frac{1}{2}\left(\frac{\partial w}{\partial x}\right)^2 - z\frac{\partial^2 w}{\partial x^2}, \\ e_{yy} &= \frac{\partial v}{\partial y} + \frac{1}{2}\left(\frac{\partial w}{\partial y}\right)^2 - z\frac{\partial^2 w}{\partial y^2}, \\ e_{xy} &= \frac{1}{2}\left(\frac{\partial u}{\partial y} + \frac{\partial v}{\partial x}\right) + \frac{\partial w}{\partial x}\frac{\partial w}{\partial y} - z\frac{\partial^2 w}{\partial x \partial y}. \end{aligned} \tag{10.2}$$

Non-zero bending-torsion symmetric tensor components, in the case of the dependence of the displacements and rotations fields take the form:

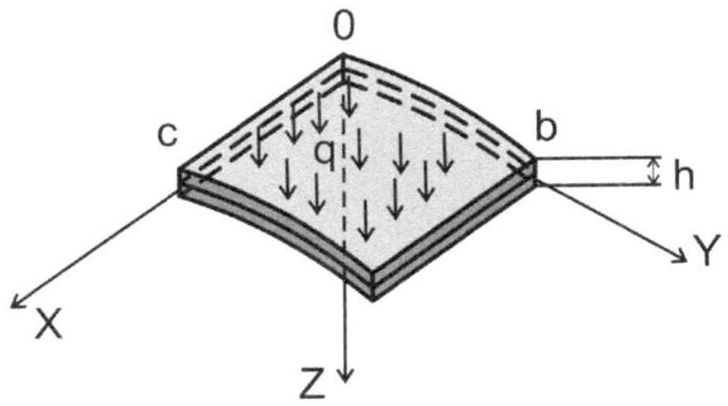

Fig. 10.1 Calculation scheme of the plate

$$\chi_{xx} = \frac{\partial^2 w}{\partial x \partial y}, \quad \chi_{yy} = -\frac{\partial^2 w}{\partial x \partial y}, \quad \chi_{xy} = \frac{1}{2}\left(\frac{\partial^2 u}{\partial y^2} - \frac{\partial^2 v}{\partial x^2}\right),$$
$$\chi_{yz} = \frac{1}{4}\left(\frac{\partial^2 v}{\partial x^2} - \frac{\partial^2 u}{\partial x \partial y}\right), \quad \chi_{xz} = \frac{1}{4}\left(\frac{\partial^2 v}{\partial x \partial y} - \frac{\partial^2 u}{\partial y^2}\right). \tag{10.3}$$

Defining relations taking into account Young's proposals and the Duhamel-Neumann's law take the form:

$$\sigma_{xx} = \frac{E}{1-\nu^2}\left[e_{xx} + \nu e_{yy}\right] - \frac{E}{1-\nu}\alpha_t\left[\Theta - \Theta_0\right],$$
$$\sigma_{yy} = \frac{E}{1-\nu^2}\left[e_{yy} + \nu e_{xx}\right] - \frac{E}{1-\nu}\alpha_t\left[\Theta - \Theta_0\right], \tag{10.4}$$
$$\sigma_{xy} = \frac{E}{1+\nu}e_{xy}, \quad m_{ij} = \frac{El^2}{1+\nu}\chi_{ij}, i, j = x, y, z,$$

σ_{ij} are stress tensor components, m_{ij} are higher-order symmetric moment tensor components, E is Young's modulus, ν is Poisson's ratio, l is an additional independent material length parameter related to the bending-torsion tensor $\boldsymbol{\chi}$, α_t is the thermal expansion coefficient, Θ is a known function of the plate absolute temperature at time $t \in (t_0, t]$ Θ_0 is the initial temperature of the plate at time t_0.

The motion equations of continuous plate, boundary and initial conditions are obtained from the Hamilton's variational principle:

$$\int_{t_0}^{t_1} \left(\delta K - \delta U + \delta W_\epsilon + \delta W_q\right) dt = 0, \tag{10.5}$$

here

$$K = \frac{1}{2}\rho \int_{\Omega} \left[\left(\frac{\partial u_x}{\partial t}\right)^2 + \left(\frac{\partial u_y}{\partial t}\right)^2 + \left(\frac{\partial u_z}{\partial t}\right)^2\right] d\Omega,$$

$$U = \int_{\Omega} \left(\sigma_{ij}\epsilon_{ij} + m_{ij}\chi_{ij}\right) d\Omega, \quad i, j = x, y, z$$

are kinetic and potential energies,

$$\delta W_q = \int_0^c \int_0^b q(x, y, t)\delta w \, dx \, dy, \quad \delta W_\epsilon = \int_{\Omega} \rho\epsilon \frac{\partial w}{\partial t}\delta w \, d\Omega,$$

are the work of external forces associated with the external environment dissipation and the work of the external forces associated with external normal distributed vibration load, ϵ is the medium dissipation factor, ρ is the plate material density, $q(x, y, t)$ is the external normal load given as $q = q_0(x, y)Sin(\omega_p t)$, where ω_p is the frequency and $q_0(x, y)$ is the vibration load amplitude.

Designations of force, moments and high order moments are introduced as follows:

$$\{N_{xx}, M_{xx}\} = \int_{-\frac{h}{2}}^{\frac{h}{2}} \sigma_{xx} z^{\{0,1\}} dz, \quad \{T, H\} = \int_{-\frac{h}{2}}^{\frac{h}{2}} \sigma_{xy} z^{\{0,1\}} dz,$$
$$Y_{xx} = \int_{-\frac{h}{2}}^{\frac{h}{2}} m_{xx}\, dz, \quad Y_{zx} = \int_{-\frac{h}{2}}^{\frac{h}{2}} m_{zx}\, dz, \quad Y_{xy} = \int_{-\frac{h}{2}}^{\frac{h}{2}} m_{xy}\, dz, \qquad x \leftrightarrows y. \tag{10.6}$$

Taking into account the designations (10.6), the motion equations of continuous plate will have the form:

$$\frac{\partial N_{xx}}{\partial x} + \frac{\partial T}{\partial y} + \frac{1}{2}\frac{\partial^2 Y_{yz}}{\partial y^2} + \frac{1}{2}\frac{\partial^2 Y_{xz}}{\partial x \partial y} = \rho h \frac{\partial^2 u}{\partial t^2};$$
$$\frac{\partial N_{yy}}{\partial y} + \frac{\partial T}{\partial x} - \frac{1}{2}\frac{\partial^2 Y_{xz}}{\partial x^2} - \frac{1}{2}\frac{\partial^2 Y_{yz}}{\partial x \partial y} = \rho h \frac{\partial^2 v}{\partial t^2};$$
$$\frac{\partial^2 M_{xx}}{\partial x^2} + \frac{\partial^2 M_{yy}}{\partial y^2} + 2\frac{\partial^2 H}{\partial x \partial y} - \frac{\partial^2 Y_{xx}}{\partial x \partial y} + \frac{\partial^2 Y_{yy}}{\partial x \partial y} + \frac{\partial^2 Y_{xy}}{\partial x^2} - \frac{\partial^2 Y_{xy}}{\partial y^2} + \tag{10.7}$$
$$+\frac{\partial}{\partial x}\left(N_{xx}\frac{\partial w}{\partial x}\right) + \frac{\partial}{\partial y}\left(N_{yy}\frac{\partial w}{\partial y}\right) + 2\frac{\partial T}{\partial x}\frac{\partial w}{\partial y} + 2\frac{\partial T}{\partial y}\frac{\partial w}{\partial x} + 4T\frac{\partial^2 w}{\partial x \partial y} +$$
$$+2q = \rho h \frac{\partial^2 w}{\partial t^2} + \epsilon \rho h \frac{\partial w}{\partial t}.$$

Initial conditions:

$$w(x, y, 0) = \xi_1(x, y), \quad \frac{\partial w(x, y, 0)}{\partial t} = \xi_2(x, y),$$
$$u(x, y, 0) = \xi_3(x, y), \quad \frac{\partial u(x, y, 0)}{\partial t} = \xi_4(x, y), \tag{10.8}$$
$$v(x, y, 0) = \xi_5(x, y), \quad \frac{\partial v(x, y, 0)}{\partial t} = \xi_6(x, y).$$

The generalized boundary conditions are written as follows:

$$
\begin{aligned}
&\delta u = 0 \text{ or } \left\{N_{xx} - \tfrac{1}{2}\tfrac{\partial Y_{zx}}{\partial y}\right\}_{\Gamma_x} = 0, \quad \left\{T - \tfrac{1}{2}\tfrac{\partial Y_{zy}}{\partial y} - \tfrac{1}{2}\tfrac{\partial Y_{zx}}{\partial x}\right\}_{\Gamma_y} = 0;\\
&\frac{\partial \delta u}{\partial x} = 0 \text{ or } \{Y_{zx}\}_{\Gamma_y} = 0; \quad \tfrac{\partial \delta u}{\partial y} = 0 \text{ or } \{Y_{zx}\}_{\Gamma_x} = 0, \quad \left\{Y_{zy}\right\}_{\Gamma_y} = 0;\\
&\delta v = 0 \text{ or } \left\{-T + \tfrac{1}{2}\tfrac{\partial Y_{zx}}{\partial x} + \tfrac{1}{2}\tfrac{\partial Y_{zy}}{\partial y}\right\}_{\Gamma_x} = 0, \quad \left\{N_{yy} - \tfrac{1}{2}\tfrac{\partial Y_{zy}}{\partial x}\right\}_{\Gamma_y} = 0;\\
&\frac{\partial \delta v}{\partial x} = 0 \text{ or } \{Y_{zx}\}_{\Gamma_x} = 0, \quad \left\{Y_{zy}\right\}_{\Gamma_y} = 0; \quad \tfrac{\partial \delta v}{\partial y} = 0 \text{ or } \left\{Y_{zy}\right\}_{\Gamma_x} = 0;\\
&\delta w = 0 \text{ or } \left\{N_{xx}\tfrac{\partial w}{\partial x} + \tfrac{\partial M_{xx}}{\partial x} + 2\tfrac{\partial H}{\partial y} + 2T\tfrac{\partial w}{\partial y} - \tfrac{\partial Y_{xx}}{\partial y} + \tfrac{\partial Y_{yy}}{\partial y} - \tfrac{\partial Y_{xy}}{\partial x}\right\}_{\Gamma_x} = 0,\\
&\left\{N_{yy}\frac{\partial w}{\partial y} + \frac{\partial M_{yy}}{\partial y} + 2\frac{\partial H}{\partial x} + 2T\frac{\partial w}{\partial x} - \frac{\partial Y_{xx}}{\partial x} + \frac{\partial Y_{yy}}{\partial x} - \frac{\partial Y_{xy}}{\partial y}\right\}_{\Gamma_y} = 0;\\
&\frac{\partial \delta w}{\partial x} = 0 \text{ or } \left\{M_{xx} - Y_{xy}\right\}_{\Gamma_x} = 0, \quad \left\{2H - Y_{xx} + Y_{yy}\right\}_{\Gamma_y} = 0;\\
&\frac{\partial \delta w}{\partial y} = 0 \text{ or } \left\{2H - Y_{xx} + Y_{yy}\right\}_{\Gamma_x} = 0, \quad \left\{M_{yy} - Y_{xy}\right\}_{\Gamma_\beta} = 0.
\end{aligned}
\tag{10.9}
$$

10.2.2 Heat Equation

To the motion equations of the continuous plate element add the three-dimensional heat equation in a coupled formulation. Cattaneo-Vernot's hypotheses, in combination with the heat balance equation, make it possible to obtain a hyperbolic equation.

$$
C\rho\tau_q\frac{\partial^2\Theta}{\partial t^2} + C\rho\frac{\partial\Theta}{\partial t} = \lambda\left(\frac{\partial^2\Theta}{\partial x^2} + \frac{\partial^2\Theta}{\partial y^2} + \frac{\partial^2\Theta}{\partial z^2}\right) - \frac{\Theta_0 E\alpha_t}{1-\nu}\left(\frac{\partial e_{xx}}{\partial t} + \frac{\partial e_{yy}}{\partial t}\right),
\tag{10.10}
$$

here C is the specific heat, λ is the thermal conductivity coefficient. The equation (10.10) should be supplemented with initial conditions:

$$
\Theta(x, y, z, 0) = f_1(x, y, z); \qquad \frac{\partial\Theta(x, y, z, 0)}{\partial t} = f_2(x, y, z).
\tag{10.11}
$$

Consider the first kind boundary conditions:

$$
\Theta_\Pi(t) = F(t).
\tag{10.12}
$$

10.2.3 Meshed Structure Accounting

Assume that the original plate consists of several families of densely spaced ribs. Each family is characterized by the width of the family ribs, distance between the family ribs, and the inclination angle of the rib relative to the abscissa axis positive direction Fig. 10.2.

The relationship between stresses and higher orders moments of the continuous plate with stresses and higher orders moments in ribs has the form:

$$\begin{gathered}\sigma_{xx}=\sum_{j=1}^{n}\frac{\sigma_x^j\delta_j Cos^2\varphi_j}{a_j},\quad \sigma_{yy}=\sum_{j=1}^{n}\frac{\sigma_x^j\delta_j Sin^2\varphi_j}{a_j},\\ \sigma_{xy}=\sum_{j=1}^{n}\frac{\sigma_x^j\delta_j Cos\varphi_j Sin\varphi_j}{a_j},\quad m_{xx}=\sum_{j=1}^{n}\frac{m_x^j\delta_j Cos^2\varphi_j}{a_j},\\ m_{yy}=\sum_{j=1}^{n}\frac{m_x^j\delta_j Sin^2\varphi_j}{a_j},\quad m_{xy}=\sum_{j=1}^{n}\frac{m_x^j\delta_j Cos\varphi_j Sin\varphi_j}{a_j},\\ m_{xz}=\sum_{j=1}^{n}\frac{m_x^j\delta_j Cos\varphi_j}{a_j},\quad m_{yz}=\sum_{j=1}^{n}\frac{m_x^j\delta_j Sin^2\varphi_j}{a_j}\end{gathered}\tag{10.13}$$

These relations are obtained from the conditions of acting equality on the same elementary areas of a plate consisting of ribs system and continuous plate equivalent to it. In expressions n is the number of ribs families in the meshed plate δ_j is the distance between the j families ribs, a_j is the j families ribs width, φ_j is the angle between the X axis and the j families ribs axis.

Additional conditions for the original meshed plate static equivalence and the equivalent continuous plate are obtained using the Lagrange multiplier method from the condition that the functional reaches stationary values. When constructing the

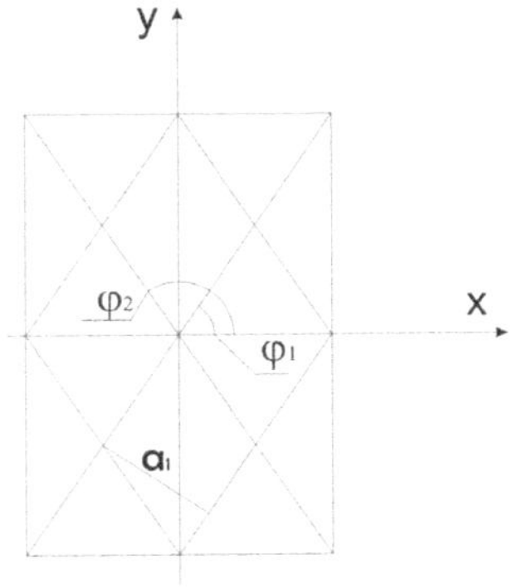

Fig. 10.2 Meshed structure of the plate

functional, expressions for the deformation potential energy were used, expressed in terms of higher orders stresses and moments.

$$\{\sigma^j, m^j\} = \{\sigma_{xx}, m_{xx}\}Cos^2\varphi_j + \{\sigma_{yy}, m_{yy}\}Sin^2\varphi_j + \{\sigma_{xy}, m_{xy}\}Cos\varphi_j Sin\varphi_j,$$
$$\{\sigma_z^j, m_z^j\} = \{\sigma_{zx}, m_{zx}\}Cos\varphi_j + \{\sigma_{zy}, m_{zy}\}Sin\varphi_j. \tag{10.14}$$

Substituting (10.14) into (10.13), taking into account (10.6) and the designation

$$A_{sk} = \sum_{j=1}^{n} \frac{\delta_j Cos^s\varphi_j Sin^k\varphi_j}{a_j} \quad s, k = \overline{0, 4},$$

we get expressions for the meshed plate forces and moments, through the forces and moments in an equivalent continuous plate. Substituting the obtained relations into the motion equations of continuous plate (10.7), we obtain the motion equations of plate, taking into account its meshed structure.

10.3 Numerical Experiment

The experiments considered the plate consisting of two mutually perpendicular ribs families $c = b = 0.001$ cm, $h = 0.002$ cm, $a_1 = a_2 = a = 0.00002$ cm, $\delta_1 = \delta_2 = \delta = 0.00002$ cm, $\varphi_1 = 45^o$, $\varphi_2 = 135^o$, $l = 0.002$ cm, $\nu = 0.37$, $\rho = 0.002267\ \frac{\text{kg}}{\text{cm}^3}$, $E = 10500060\ \frac{\text{kgf}}{\text{cm}^2}$.

We consider the plate ends to be rigid clamped. Nonlinear differential problems in partial derivatives for the plate element motion equations and the heat equation are reduced to Cauchy problems by the finite differences method of the second accuracy order. These differential problems have different dimensions, therefore, at each time step, the temperature moments and forces in the middle surface were found by the Simpson's method. The obtained values of Mt and Nt were inserted as a load into the Cauchy problem for the plate element motion equations. Both Cauchy problems were solved by the Runge-Kutta method of the fourth accuracy order. The choice rationale of numerical methods in the study of micropolar meshed cylindrical panels nonlinear oscillations is given in the article [23]. The statics problem was previously solved to obtain the deflection, provided that the plate was in a stationary temperature field. The resulting deflection was taken as the initial conditions for the plate motion equation. The system of nonlinear differential equations in partial derivatives without taking into account dynamic terms was reduced to the nonlinear algebraic equations system by the finite differences method of the second accuracy order, which was solved by Newton's method.

The need to take into account temperature effects in the dynamics analysis of the considered plates is confirmed by the data in Figs. 10.3 and 10.4. The signals obtained taking into account the stationary temperature field $\Theta = 300$ K (20 °C) in the model

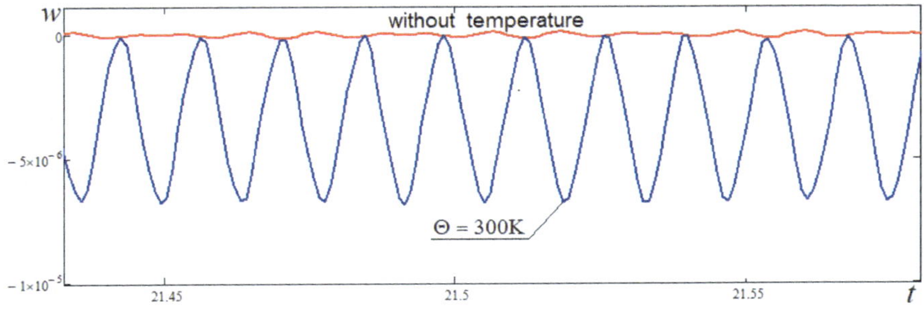

Fig. 10.3 Signal without temperature and $\Theta = 300$ K

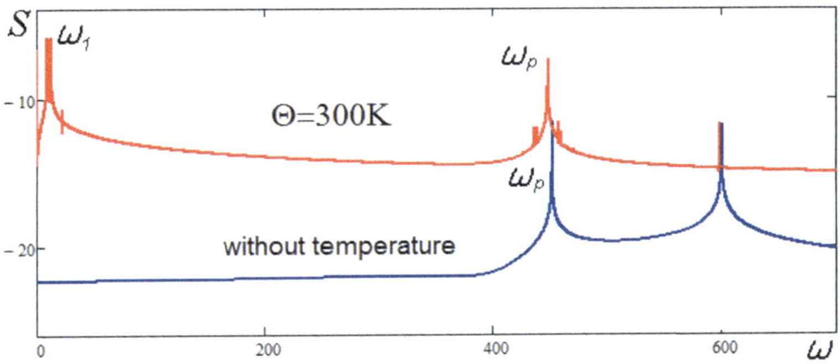

Fig. 10.4 Fourier spectrum without temperature and $\Theta = 300$ K

differ significantly from the signals obtained without taking into account the temperature effect Fig. 10.3. The signals frequency content is also different Fig. 10.4. In the Fourier spectum the S axis is the intensity of the signal harmonics on a logarithmic scale in decibels (dB), the ω axis is the oscillation frequency. The plate was subjected to the normal distributed load with the frequency close to the plate natural frequency under consideration $\omega_p = 450$ Hz, $q_0 = 0.01\ \frac{\text{kgf}}{\text{cm}^2}$.

The Figures 10.5 and 10.6 show the Fourier spectum and signals in the stationary temperature field case and the case of non-stationary temperature field, taking into account self-heating $\Theta = 600$ K. In the stationary temperature field case, the spectrum contains the excitation frequency and the natural oscillations frequency, as well as the high frequency $\omega_2 = 1369$ Hz. In the non-stationary temperature field case, taking into account self-heating, there is still the incommensurable frequency $\omega_1 = 11.683$ Hz, the high frequency $\omega_3 = 1580.4$ Hz. The signals for these two cases are shown in Fig. 10.6. Parameters of this experiment $\omega_p = 450$ Hz, $q_0 = 1\ \frac{\text{kgf}}{\text{cm}^2}$.

The data in Table 10.1 show that even with sufficiently large plate sizes, large thickness of the 10^{-3} order, short time range and low temperature, self-heating increases the plate deflection. Parameters of this experiment $\omega_p = 450$ Hz, $q_0 = 20\ \frac{\text{kgf}}{\text{cm}^2}$, $\Theta = 300$ K.

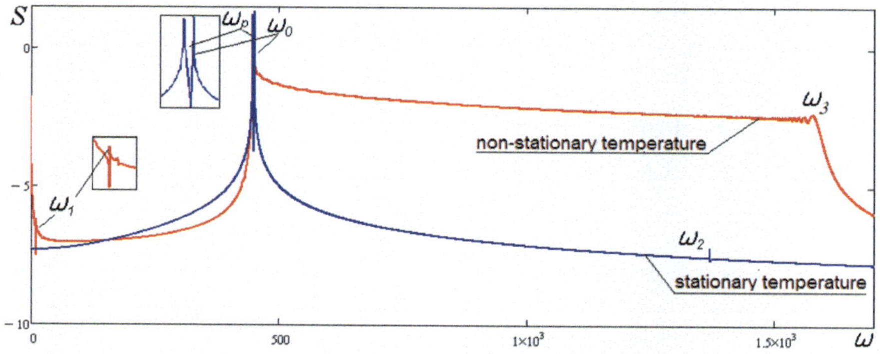

Fig. 10.5 Fourier spectrum stationary temperature and non-stationary temperature with self-heating $\Theta = 600$ K

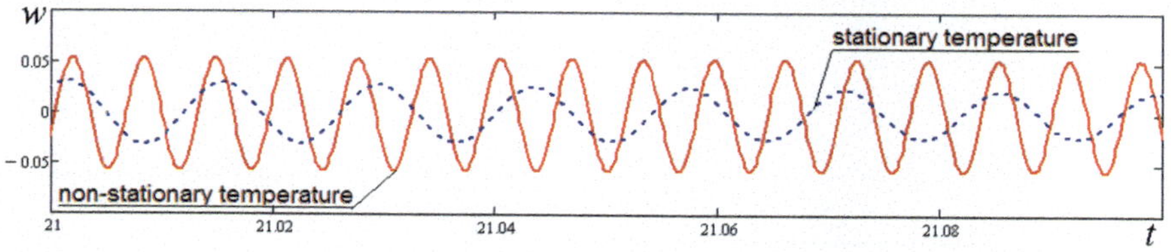

Fig. 10.6 Signal stationary temperature and non-stationary temperature with self-heating $\Theta = 600$ K

Table 10.1 Influence of self-heating on deflection at $\Theta = 300$ K

Time	w without self-heating (cm)	w with self-heating (cm)	Difference (%)
<= 2 s	−0.0189513	−0.0189513	0
10 s	−0.0112791	−0.0112792	0.001
60 s	−0.00625718	−0.00625873	0.025

Conclusion

In this work, a mathematical model of nonlinear vibrations of a mesh plate was constructed based on a modified couple stress theory. In this model is taking into account temperature effects.

The study showed that taking into account temperature influences does not change the scenario for the oscillations transition from harmonic to chaotic, however, there is a difference in the oscillations character under other identical experimental conditions, taking into account temperature influences and without taking them into account.Heating leads to a change in the plate equilibrium state and an increase in the amplitude of its oscillations. Self-heating leads to a change in the frequency characteristics of the signal, in particular to the appearance of subharmonics both in the excitation frequency region and in the low frequency region.

Thus, when studying the micro and nanoscale meshed plates nonlinear dynamics, it is necessary to take into account temperature effects and the self-heating effect.

The work was carried out with the support of the RSF No 22-21-00331 https://rscf.ru/project/22-21-00331/

References

1. Chuan MW, Riyadi MA, Hamzah A et al (2021) Impact of phonon scattering mechanisms on the performance of silicene nanoribbon field-effect transistors. Results Phys 29:104714. https://doi.org/10.1016/j.rinp.2021.104714
2. Wong KL, Chuan MW, Afiq Hamzah A, Rusli S et al (2020) Performance metrics of current transport in pristine graphene nanoribbon field-effect transistors using recursive non-equilibrium Green's function approach. Superlattices Microstruct 145:106624. https://doi.org/10.1016/j.spmi.2020.106624
3. Altenbach H, Eremeyev VA (2009) On the linear theory of micropolar plates. ZAMM J Appl Math Mech 89(4)"242–256. https://doi.org/10.1002/zamm.200800207
4. Eremeev VA, Zubov LM (2008) Mechanics of elastic shells. Nauka, Moscow
5. Norouzzadeh A, Ansari R, Darvizeh M (2021) Isogeometric dynamic analysis of shells based on the nonlinear micropolar theory. Int J Non-Linear Mech 135:103750. https://doi.org/10.1016/j.ijnonlinmec.2021.103750
6. Carrera E, Zozulya VV (2022) Carrera unified formulation (CUF) for the micropolar plates and shells. I. Higher order theory. Mech Adv Mater Struct 29(6):773–795. https://doi.org/10.1080/15376494.2020.1793241
7. Sargsyan A, Sargsyan S (2021) Geometrically nonlinear models of static deformation of micropolar elastic thin plates and shallow shells. ZAMM Zeitschrift fur Angewandte Mathematik und Mechanik 101(5):e202000148. https://doi.org/10.1002/zamm.202000148
8. Zubov LM, Kolesnikov AM, Rudenko OV (2019) Exact solutions of nonlinear micropolar elastic theory for compressible solids. Adv Struct Mater 110:771–798. https://doi.org/10.1007/978-3-030-17747-8_37
9. Krylova EY, Papkova IV, Krysko VA (2020) Mathematical modeling of complex oscillations of flexible micropolar mesh cylindrical panels. Russ Phys J 62:1649–1655. https://doi.org/10.1007/s11182-020-01887-x
10. Mazur O, Kurpa L, Awrejcewicz J (2020) Vibrations and buckling of orthotropic small-scale plates with complex shape based on modified couple stress theory. ZAMM Zeitschrift fur Angewandte Mathematik und Mechanik 100(11):e202000009. https://doi.org/10.1002/zamm.202000009
11. Yang F, Chong ACM, Lam DCC, Tong P (2002) Couple stress based strain gradient theory for elasticity. Int J Solids Struct 39:2731–2743. https://doi.org/10.1016/S0020-7683(02)00152-X
12. Krysko VA, Awrejcewicz J, Papkova IV, Krysko VA (2021) Chaotic vibrations of size-dependent flexible rectangular plates. Chaos 31(4):043119. https://doi.org/10.1063/5.0044630
13. Krysko VA, Awrejcewicz J, Papkova IV, Saltykova OA, Krysko AV (2019) Chaotic contact dynamics of two microbeams under various kinematic hypotheses. Int J Nonlinear Sci Numer Simul 20:373–386. https://doi.org/10.1515/ijnsns-2018-0132
14. Krysko VA, Awrejcewicz J, Krylova EY, Papkova IV, Krysko AV (2018) Non-symmetric forms of non-linear vibrations of flexible cylindrical panels and plates under longitudinal load and additive white noise. J Sound Vib 423:212–229. https://doi.org/10.1016/j.jsv.2018.02.065

15. Awrejcewicz J, Krysko VA, Zhigalov MV, Papkova IV, Krysko VA (2018) Mathematical models for quantifying flexible multilayer orthotropic shells under transverse shear stresses. Compos Struct 204:896–911. https://doi.org/10.1016/j.compstruct.2018.07.052
16. Xu FFY, Sahmani S, Safaei B (2020) Modified couple stress-based geometrically nonlinear oscillations of porous functionally graded microplates using NURBS-based isogeometric approach. Comput Methods Appl Mech Eng 372:113400. https://doi.org/10.1016/j.cma.2020.113400
17. Ghobadi A, Beni YT, Żur KK (2021) Porosity distribution effect on stress, electric field and nonlinear vibration of functionally graded nanostructures with direct and inverse flexoelectric phenomenon. Compos Struct 259:113220. https://doi.org/10.1016/j.compstruct.2020.113220
18. Dell'Isola F, Steigman D (2015) A Two-dimensional gradient-elasticity theory for woven fabrics. J Elast 118(1):113–125. https://doi.org/10.1007/s10659-014-9478-1
19. Eremeev V (2018) A On a nonlinear model of the mesh shell. Izvestiya of the Russian Academy of Sciences. Solid Mech 4:127–133. https://doi.org/10.3103/S002565441804012X
20. Krylova EY, Papkova IV, Yakovleva TV, Krysko VA (2019) Theory of vibrations of carbon nanotubes like flexible micropolar mesh cylindrical shells taking into account shift. Izv Saratov Univ (N. S.) Ser Math Mech Inform 19(3):305–316. https://doi.org/10.18500/1816-9791-2019-19-3-305-316
21. Cattaneo C (1958) Sur une forme de l'equation de la chaleur eliminant le paradoxe d'une propagation instantanee. Comptes Rendus de l'Académie des Sci 247:431
22. Vernotte P (1958) Les paradoxes de la theorie continue de l'equation de la chaleur. Instantanee 246:3154
23. Krylova EY, Papkova IV, Sinichkina OA (2020) Justification of the choice of numerical methods in the study of nonlinear micropolar mesh cylindrical panel's oscillations. In: IOP conference series: materials science and engineering international conference of young scientists and students "topical problems of mechanical engineering", ToPME 2019, p 012118. https://doi.org/10.1088/1757-899X/747/1/012118

Chapter 11
Nonlinear Deformation of Plane with Rigid Elliptical Inclusion Loaded by Force and Moment

Yulia Malkova

Abstract An exact analytical solution is obtained for a nonlinear problem of elasticity theory for a plane with a rigid elliptical inclusion. A concentrated force and a moment are applied at the center of inclusion. The elastic properties of the plane are modeled by John's harmonic material. Nominal stresses and displacements are expressed in terms of two analytical functions of a complex variable, which are determined from the boundary conditions on the contour of inclusion. Calculations of nominal stresses on the contour joining the plane with inclusion are performed.

Keywords Nonlinear plane problem · Rigid elliptical inclusion · John's harmonic material · Concentrated force and moment

11.1 Introduction

Model of F. John's harmonic material [1] makes it possible to investigate large deformations of real materials, in particular elastomers. In the case of small deformations, the model leads to Hooke's law. The model was significantly developed and applied in works [2–7] and others. Theoretical solutions to some problems, for example, the stretching of rubber sheets with circular or elliptical holes, have been experimentally confirmed [4]. A complex formulation of the equations of nonlinear plane problems is given in [4], where plane deformation and plane stress state for a plate with an elliptical hole were studied. Further development of the complex method is contained in work [5] for an inhomogeneous plate with an interfacial crack. However, the solution was completed only for a homogeneous plate with a crack. Complex equations of nonlinear plane problems for F. John's material, which differ from those previously known, are given in [6]. In the same work, a method for solving equations based on the introduction of functions of stress jumps and displacements at the interface between two materials is proposed. Using this method, exact solutions of problems for a crack and a concentrated force at the interface are obtained. The study of the

Y. Malkova (✉)
Saint Petersburg State University, Universitetskaya Emb. 7-9, St. Petersburg, Russia
e-mail: y.malkova@spbu.ru

H. Altenbach and V. Kuzkin (eds.), *Advanced Problems in Mechanics*,
https://doi.org/10.1007/978-3-031-80103-7_11

interfacial crack problem was continued in [7] for the case of uniform pressure on the crack faces. It is shown for the first time that there are critical pressures, the excess of which leads to the loss of stability of the material and large stresses and displacements in the vicinity of the crack. In a number of works, for example [8, 9] the nonlinear problem of elliptical inclusion in a plane was studied. It is found that the tensor of nominal stresses is constant in the inclusion at given constant stresses at infinity.

11.2 Nonlinear Equations of Plane Deformation for John's Harmonic Material

To solve the nonlinear elasticity problem, we use the equations of equilibrium and compatibility of the reference configuration (there are no body forces) [6]

$$\mathtt{div}\,\mathbf{S} = \mathbf{0}, \quad \mathbf{rot}\,\mathbf{G}^T = \mathbf{0}, \tag{11.1}$$

where $\mathbf{S} = s_{\alpha\beta}\,\mathbf{e}_\alpha \mathbf{e}_\beta$ is tensor of nominal (Piola) stresses, $\mathbf{G} = g_{\alpha\beta}\,\mathbf{e}_\alpha \mathbf{e}_\beta$ is deformation gradient, $\mathbf{e}_i$ is orthonormal vector basis of Cartesian coordinates of the reference configuration.

Let us pass in (11.1) to the components of the tensors and write in the complex form for the plane deformation

$$(s_{11} + is_{12})'_1 + i(s_{22} - is_{21})'_2 = 0, \quad (g_{22} - ig_{12})'_1 + i(g_{11} + ig_{21})'_2 = 0, \tag{11.2}$$

prime denote partial derivatives in Cartesian coordinates (x_1, x_2).

The equations (11.2) are satisfied identically by substituting of expressions

$$s_{11} + is_{12} = \frac{\partial \sigma}{\partial z} - \frac{\partial \sigma}{\partial \overline{z}}, \quad s_{22} - is_{21} = \frac{\partial \sigma}{\partial z} + \frac{\partial \sigma}{\partial \overline{z}}, \tag{11.3}$$

$$g_{11} + ig_{21} = \frac{\partial \zeta}{\partial z} + \frac{\partial \zeta}{\partial \overline{z}}, \quad g_{22} - ig_{12} = \frac{\partial \zeta}{\partial z} - \frac{\partial \zeta}{\partial \overline{z}},$$

where $z = x_1 + ix_2$, $\zeta = \zeta_1 + i\zeta_2$ are the complex variables of the initial and current configurations, $\sigma = f_1 + if_2$ is the main force vector on the arc in the z plane [6], f_1, f_2 are projections onto the coordinate axes of the force vector. The complex functions $\zeta(z, \overline{z})$ and $\sigma(z, \overline{z})$ are found from the law of elasticity and the boundary conditions of the problem.

John's harmonic material model is defined by the strain-energy density (elastic potential) having the form [1]

$$\Phi = 2\mu[F(I) - J], \quad I = \lambda_1 + \lambda_2, \quad J = \lambda_1\lambda_2. \tag{11.4}$$

Here λ_1, λ_2 are principal stretches, $F(I)$ is function [4, 5]

$$2\mu F'(I) = \frac{1}{4b}[I + \sqrt{I^2 - 16bc}\,].$$

The constants b and c are uniquely determined by transforming the elasticity law into Hooke's law at small deformations

$$4\mu b = 1 + \frac{\mu}{\lambda + 2\mu}, \quad c = \mu\left(1 - \frac{\mu}{\lambda + 2\mu}\right),$$

λ, μ are the Lame constants. A review of the main works where the John's material model was used to solve nonlinear elasticity problems is given in [9].

The invariants I and J are expressed in terms of the deformation gradient components as follows

$$I = \lambda_1 + \lambda_2 = \sqrt{(g_{11} + g_{22})^2 + (g_{21} - g_{12})^2} = 2\left|\frac{\partial \zeta}{\partial z}\right|,$$

$$J = \lambda_1 \lambda_2 = g_{11}g_{22} - g_{12}g_{21}.$$

These equalities are derived from the polar expansion of the deformation gradient $\mathbf{G} = \mathbf{Q} \cdot \Lambda$, where $\mathbf{Q}$ is the orthogonal tensor, Λ is tensor of multiplicity of elongations.

The elasticity law for the components of the nominal stress tensor is obtained from the potential (11.4) according to the formula $s_{ij} = \partial\Phi/\partial g_{ji}$ [6].

Substituting these stresses into (11.3), we obtain two equations for the functions ζ and σ, from which we find [3]

$$\zeta = b\,\varphi(z) + \overline{\psi(z)} + \frac{cz}{\overline{\varphi'(z)}}, \quad \sigma = (1 - 2\mu b)\,\varphi(z) - 2\mu\overline{\psi(z)} - 2\mu\frac{cz}{\overline{\varphi'(z)}}. \tag{11.5}$$

Analytic functions $\varphi(z)$ and $\psi(z)$ are determined from boundary value problems on the inclusion contour.

Substituting functions (11.5) into formulas (11.3), we write the stresses and deformations in terms of the functions $\varphi(z)$ and $\psi(z)$

$$s_{11} + is_{12} = (1 - 2\mu b)\varphi'(z) - 2\mu\left[\frac{c}{\overline{\varphi'(z)}} - \overline{\psi'(z)} + \frac{cz\overline{\varphi''(z)}}{\overline{\varphi'^2(z)}}\right],$$

$$s_{22} - is_{21} = (1 - 2\mu b)\varphi'(z) - 2\mu\left[\frac{c}{\overline{\varphi'(z)}} + \overline{\psi'(z)} - \frac{cz\overline{\varphi''(z)}}{\overline{\varphi'^2(z)}}\right], \tag{11.6}$$

$$g_{11} + i g_{21} = b\varphi'(z) + \frac{c}{\overline{\varphi'(z)}} + \overline{\psi'(z)} - \frac{cz\overline{\varphi''(z)}}{\overline{\varphi'^2(z)}},$$

$$g_{22} - i g_{12} = b\varphi'(z) + \frac{c}{\overline{\varphi'(z)}} - \overline{\psi'(z)} + \frac{cz\overline{\varphi''(z)}}{\overline{\varphi'^2(z)}}.$$

11.3 Solution of the Second Fundamental Problem

An infinite plane containing a rigid elliptical inclusion is considered. A concentrated force and a moment are applied at the center of the ellipse. Under the action of an external force and moment, the elliptical inclusion obtains a translational displacement z^0 and a rotation about the center by an angle α. The parameters z^0 and α are unknown, they are determined in terms of the given external force and moment. Disposing of the translational movement of the entire plane, we assume that the origin of Cartesian coordinates is always located in the center of the ellipse, i.e. parameter $z^0 = 0$. At infinity, the stresses are zero: $s_{ij} \to 0$ for $|z| \to \infty$.

The boundary condition is the current coordinates of the contour L of the elliptic inclusion, which are known up to a constant α,

$$\zeta = b\varphi(z) + \overline{\psi(z)} + \frac{cz}{\overline{\varphi'(z)}} = g(z) = z\,e^{i\alpha}, \quad z \in L. \tag{11.7}$$

Thus, it is necessary to solve the second boundary value problem of the nonlinear theory of elasticity (11.7) for a plane with an elliptical hole, the edge of which is displaced as a rigid whole.

The principal vector and principal moment of stresses $\mathbf{s}_n = \mathbf{n} \cdot \mathbf{S}$ on any closed loop containing an elliptical inclusion must balance the external force and moment

$$F_1 + iF_2 = \oint (s_{n1} + is_{n2})\,ds; \quad M = \oint (x_1 s_{n2} - x_2 s_{n1})\,ds. \tag{11.8}$$

The function ζ is subject to the condition of single-valuedness on the indicated closed loop.

To find a solution of Eq. (11.7), we first perform a conformal mapping of a plane with an elliptic inclusion onto a plane with a circular inclusion of unit radius [10]

$$z = \eta(\xi) = R\left(\xi + m\,\xi^{-1}\right); \quad R > 0, \; 0 \le m < 1. \tag{11.9}$$

We set $\xi = re^{i\theta}$, the parameters (r, θ) are the polar coordinates of the ξ plane and the curvilinear coordinates of the z plane. Semi-axes of the ellipse are $a_0 = R(1 + m)$, $b_0 = R(1 - m)$.

In the complex functions $\varphi(z)$ and $\psi(z)$, we pass to the variable ξ. Formulas (11.5) take the form

$$\zeta = b\varphi(\xi) + \overline{\psi(\xi)} + c\frac{\eta(\xi)\overline{\eta'(\xi)}}{\overline{\varphi'(\xi)}},$$

$$\sigma = (1 - 2\mu b)\,\varphi(\xi) - 2\mu\overline{\psi(\xi)} - 2\mu c\,\frac{\eta(\xi)\overline{\eta'(\xi)}}{\overline{\varphi'(\xi)}}.$$

We write the nominal stresses and components of the deformation gradient (11.6) in curvilinear coordinates

$$s_{rr} + i s_{r\theta} = \frac{1}{\eta'(\xi)}\left(\frac{\partial\sigma}{\partial\xi} - \frac{\partial\sigma}{\partial\overline{\xi}}\,e^{-2i\theta}\right),\quad s_{\theta\theta} - i s_{\theta r} = \frac{1}{\eta'(\xi)}\left(\frac{\partial\sigma}{\partial\xi} + \frac{\partial\sigma}{\partial\overline{\xi}}\,e^{-2i\theta}\right),$$
$$g_{rr} + i g_{\theta r} = \frac{1}{\eta'(\xi)}\left(\frac{\partial\zeta}{\partial\xi} + \frac{\partial\zeta}{\partial\overline{\xi}}\,e^{-2i\theta}\right),\quad g_{\theta\theta} - i g_{r\theta} = \frac{1}{\eta'(\xi)}\left(\frac{\partial\zeta}{\partial\xi} - \frac{\partial\zeta}{\partial\overline{\xi}}\,e^{-2i\theta}\right). \qquad (11.10)$$

The boundary condition (11.7) on the hole contour is transformed to the variable $\xi = t = e^{i\theta}$ using mapping (11.9)

$$b\,\varphi(t) + \overline{\psi(t)} + c\,\frac{\eta(t)\overline{\eta'(t)}}{\overline{\varphi'(t)}} = g(t) = R\left(t + m\frac{1}{t}\right)e^{i\alpha}. \qquad (11.11)$$

Equation (11.11) allows us to find both functions $\varphi(\xi)$ and $\psi(\xi)$, similar to the linear problem [10]. Let us distinguish in these functions the non-holomorphic and holomorphic terms in the infinite domain $|\xi| > 1$

$$\varphi(\xi) = A\ln\xi + A'R\xi + \varphi_0(\xi),\quad \psi(\xi) = B\ln\xi + B'R\xi + \psi_0(\xi). \qquad (11.12)$$

The complex constants A and B are expressed in terms of the principal vector of external forces, and the constants A' and B' determine the stress distribution at infinity. The functions $\varphi_0(\xi)$ and $\psi_0(\xi)$ are holomorphic in the infinite domain. It is easy to check that the zero stress at infinity gives the values of the constants $A' = 2\mu$, $B' = 0$.

Equation (11.11) contains the boundary value of the function $\eta(\xi)\overline{\eta'(\xi)}/\overline{\varphi'(\xi)}$, which is not holomorphic either in the domain $|\xi| < 1$, nor in $|\xi| > 1$. Let us transform this function by selecting the holomorphic part in it in the domain $|\xi| < 1$. The functions $\varphi(\xi)$, $\psi(\xi)$ in Eq. (11.11) are replaced by expressions (11.12). As a result, the boundary value problem for the holomorphic functions $\varphi_0(\xi)$ and $\psi_0(\xi)$ is

$$b\,\varphi_0(t) + \overline{\psi_0(t)} + c\,\overline{\phi(t)} = g_0(t), \qquad (11.13)$$

$$g_0(t) = g(t) - (bA - \overline{B})\ln t - 2\mu bRt - cmR^2\frac{1}{2\mu R}\frac{1}{t},\quad t = e^{i\theta}.$$

Equation (11.13) contains the boundary value of the function

$$\overline{\phi(\xi)} = \frac{\eta(\xi)\,\overline{\eta'(\xi)}}{\overline{A}\xi + 2\mu R + \overline{\varphi_0'(\xi)}} - mR^2 \frac{1}{2\mu R}\frac{1}{\xi}, \tag{11.14}$$

which is holomorphic in $|\xi| < 1$. The Cauchy-type integral of this function in the domain $|\xi| > 1$ is equal to zero.

Solution of boundary value problem (11.13) has the form

$$b\,\varphi_0(\xi) = me^{i\alpha} R\frac{1}{\xi} + (bA - \overline{B}) \ln \frac{1-\xi}{\xi} - cmR^2 \frac{1}{2\mu R}\frac{1}{\xi}, \tag{11.15}$$

$$\psi_0(\xi) = e^{-i\alpha} R\frac{1}{\xi} - (b\overline{A} - B) \ln \frac{\xi - 1}{\xi} - 2\mu b R\frac{1}{\xi} - c\phi(\xi) + \psi_0(\infty).$$

The constant $\psi_0(\infty)$ does not affect the value of stresses and can be omitted. The value of this constant can be found by the same method as in [10]. In the second formula (11.15), the function

$$\phi(\xi) = \frac{\overline{\eta(\xi)}\,\eta'(\xi)}{A/\xi + 2\mu R + \varphi_0'(\xi)} - mR^2 \frac{1}{2\mu R}\xi$$

is conjugate to the function (11.14), it is holomorphic for $|\xi| > 1$: $\phi(\xi) = O(\xi^{-2})$, $|\xi| \to \infty$.

According to formulas (11.12), the functions $\varphi_0(\xi)$ and $\psi_0(\xi)$ must be holomorphic in the domain $|\xi| > 1$, but formulas (11.15) contain non-holomorphic terms. These terms disappear from the condition of periodicity of displacements on the inclusion contour.

Formulas (11.12), (11.15) together give the complete solution of the second fundamental nonlinear problem for a plane with an elliptical hole.

On the rigidly reinforced edge of the hole, we have $g_{\theta\theta} - ig_{r\theta} = 1$, using the formula

$$s_{rr} + is_{r\theta} + 2\mu(g_{\theta\theta} - ig_{r\theta}) = \varphi'(z) = \frac{\varphi'(\xi)}{\eta'(\xi)}$$

stresses $s_{rr} + is_{r\theta}$ are found on this contour

$$s_{rr} + is_{r\theta} = \varphi'(z) - 2\mu = \frac{\varphi'(\xi)}{\eta'(\xi)} - 2\mu.$$

Using formulas (11.10), we obtain the stresses

$$s_{\theta\theta} - is_{\theta r} = 2(1 - 2\mu b)\frac{\varphi'(\xi)}{\eta'(\xi)} - 4\mu c \frac{\overline{\eta'(\xi)}}{\overline{\varphi'(\xi)}} - s_{rr} - is_{r\theta}.$$

11.4 Action of Concentrated Force on an Elliptical Inclusion

Consider the problem of the action of a concentrated force on an elliptical inclusion in the absence of a concentrated moment and the stresses at infinity.

Functions (11.12) are found up to two constants A and B. To determine these constants, we use the first formula (11.8) and the condition of uniqueness of displacements on the circle contour $|\xi| = 1$. We get two equations

$$F_1 + iF_2 = -2\pi[(1 - 2\mu b)A + 2\mu\overline{B}], \quad bA - \overline{B} = 0,$$

and we find

$$A = -\frac{F_1 + iF_2}{2\pi}, \quad B = b\frac{F_1 - iF_2}{2\pi}.$$

As regards the rigid rotation of the inclusion, we can consider the inclusion to be fixed (the inclusion does not rotate) by directing the translational movement of the entire system. The functions $\varphi(\xi)$ and $\psi(\xi)$ have the form

$$\varphi(\xi) = A\ln\xi + 2\mu R\xi + \frac{mR}{b\xi}\left(1 - \frac{c}{2\mu}\right),$$

$$\psi(\xi) = B\ln\xi - 2\mu b\frac{R}{\xi} - c\phi(\xi) + \psi_0(\infty).$$

The formulas for stresses have the form

$$s_{rr} + is_{r\theta} = -\frac{F_1 + iF_2}{2\pi R}\frac{\xi}{\xi^2 - m}.$$

The same formula for these stresses takes place in the linear problem [10].

Calculations of nominal stresses on the contour of the ellipse are performed. Parameters of elliptical hole are $a_0 = 3$ cm, $b_0 = 1$ cm, $R = 2$ cm, $m = 0.5$. Plate material parameters are $\mu = 10$ kg/cm^2, $\nu = 0.4902$ and uniaxial tension $F_1 = 10$ kg/cm (a) and $F_2 = 10$ kg/cm (b) is considered. As can be seen, under stretching along the $0x$ axis, the maximum values of radial stresses are reached at the point $\theta = \pi$, and the minimum values are obtained at $\theta = 0$. Under tension along the $0y$ axis, we have a similar situation for shear stresses (Fig. 11.1).

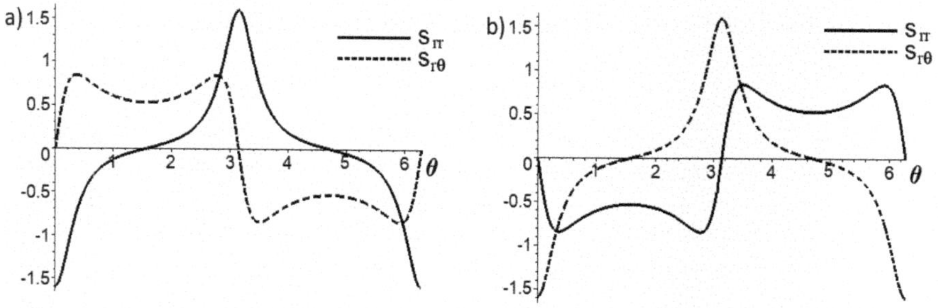

Fig. 11.1 Stresses on the contour of the inclusion

11.5 Action of Concentrated Moment on an Elliptical Inclusion

In the absence of a concentrated force, the constants $A = B = 0$, the zero stress at infinity gives $A' = 2\mu$, $B' = 0$. We also assume that due to the translational movement of the plane, the center of the ellipse is at the origin, i.e. parameter $z^0 = 0$. Equation (11.13) takes the form

$$b\,\varphi_0(t) + \overline{\psi_0(t)} + c\,\overline{\phi(t)} = \eta(t)\,e^{i\alpha} - 2\mu Rbt. \tag{11.16}$$

The function $\overline{\phi(t)}$ is the boundary value of the function

$$\overline{\phi(\xi)} = \frac{\eta(\xi)\overline{\eta'(\xi)}}{2\mu R + \overline{\varphi_0'(\xi)}}.$$

It is holomorphic for $|\xi| < 1$, except for the point $\xi = 0$, where it has a pole with principal part $mR/(2\mu\xi) + O(\xi)$.

We solve Eq. (11.16), then find the functions $\varphi(\xi)$ and $\psi(\xi)$

$$\varphi(\xi) = 2\mu R\xi + \varphi_0(\xi) = 2\mu R\xi + \frac{m}{b}R\left(e^{i\alpha} - \frac{c}{2\mu}\right)\frac{1}{\xi},$$

$$\psi(\xi) = \psi_0(\xi) = -2\mu bR\frac{1}{\xi} - c\phi(\xi) + c\frac{m}{2\mu}R\xi + R\frac{1}{\xi}e^{-i\alpha} + \psi_0(\infty).$$

Function

$$\phi(\xi) = \frac{\overline{\eta(\xi)}\,\eta'(\xi)}{2\mu R + \varphi_0'(\xi)}$$

is holomorphic in the domain $|\xi| > 1$, except for the point $\xi = \infty$, where it has the form $mR\xi/(2\mu) + O(1/\xi)$.

To determine the constant α, we use the second formula (11.8) and the condition for the uniqueness of displacements on the ellipse contour. We perform integration by parts in the integral for the moment (11.8) using the formulas $df_1 = -s_{n2}ds$, $df_2 = s_{n1}ds$:

$$M = \oint (x_1 s_{n2} - x_2 x_{n1})\, ds = -\left[x_1 f_1 + x_2 f_2\right]_L + \oint (f_1 dx_1 + f_2 dx_2).$$

If the principal force vector applied to the contour L is equal to zero, then the functions f_1 and f_2 are continuous on L, and therefore $[x_1 f_1 + x_2 f_2]_L = 0$.

Using equality

$$f_1 dx_1 + f_2 dx_2 = \frac{1}{2}(\sigma d\bar{z} + \bar{\sigma} dz)$$

and formulas (11.6), we finally obtain

$$M = \frac{1}{2}\oint (\sigma\, d\bar{z} + \bar{\sigma}\, dz) = -\mu \oint (\zeta\, d\bar{z} + \bar{\zeta}\, dz).$$

We transform the integral in this formula

$$\oint (\sigma\, d\bar{z} + \bar{\sigma}\, dz) = (1 - 2\mu b)\left[\bar{z}\varphi(z) + z\overline{\varphi(z)}\right]_L - 2\mu \oint \left[\psi(z)dz + \overline{\psi(z)}d\bar{z}\right] -$$

$$-2\mu c \oint \frac{z d\bar{z}}{\overline{\varphi'(z)}} - 2\mu c \oint \frac{\bar{z} dz}{\varphi'(z)}.$$

Due to the continuity of the functions in the first and last two terms on the hole contour, these terms are equal to zero. As a result, we obtain the following formula for the moment

$$M = -\mu \oint \left[\psi(z)dz + \overline{\psi(z)}d\bar{z}\right] = 4\pi\mu R^2 \left(1 - m^2 \frac{1 - 2\mu b}{2\mu b}\right)(1 - \cos\alpha).$$

11.6 Conclusion

The exact solution of the nonlinear problem of deformation of a plane with an elliptical inclusion under the action of a concentrated force and a moment is obtained in the work. John's harmonic material model is used to describe the material of the plane. The problems of the action of the concentrated force and moment are considered separately. Under the action of a concentrated force, the formulas for the stresses $s_{rr} + i s_{r\theta}$ of the linear and nonlinear problems coincide.

In the problem of the action of force displacements do not remain bounded at infinity. For them to be limited, it is necessary to fulfill a number of conditions:

(1) the main vector of all external forces on the boundary of the domain vanishes;
(2) stresses are zero at infinity;
(3) the infinitely remote part of the plane does not experience rotation.

Note that even if the stresses vanish at infinity and there is no rotation, the displacements still increase as $\ln r$ as $r \to \infty$ if the principal vector of external forces at the boundary is not equal to zero.

References

1. John F (1960) Plane strain problems for a perfectly elastic material of harmonic type. Commun Pure Appl Math 2(13):239–296
2. Malkov VM (2010) Introduction to non-linear elasticity. St. Petersburg State University, St. Petersburg
3. Mal'kov VM, Mal'kova YV (2017) Modeling Nonlinear Deformation of a Plate with an Elliptic Inclusion by John's Harmonic Material. Vestnik St. Petersburg Univ Math 1(50):74–81
4. Varley E, Cumberbatch E (1980) Finite deformation of elastic materials surrounding cylindrical holes. J Elast 4(10):341–405
5. Ru CQ (2002) On complex-variable formulation for finite plane elastostatics of harmonic materials. Acta Mechanica 156(3–4):219–234
6. Mal'kov VM, Mal'kova YV (2008) Plane problem of non-linear elasticity for harmonic material, Vestnik St. Peterburg Universit Ser 1 Math Mech Astron 3:114–126 (2008)
7. Malkov VM, Malkova YV, Stepanova VA (2013) Bi-material plane of John's material with interface crack loaded by pressure. Vestnic St. Petersburg State Univ Ser 1 Math Mech Astron 3:113–125 (2013)
8. Ru CQ, Schiavone P, Sudak LJ, Mioduchowski A (2005) Uniformity of stresses inside an elliptic inclusion in finite plane elastostatics. Intern J Non-Liner Mech 340(2–3):281–287
9. Mal'kov VM, Mal'kova YV (2015) Deformation of a plate with elastic elliptic inclusion. Vestnik St. Petersburg Univ Ser 1 Math Mech Astron 4:617–632
10. Muskhelishvili NI (1977) Some basic problems of the mathematical theory of elasticity. Kluwer, Leyden

Chapter 12
Influence of Heat Transfer on the Hydrodynamic Parameters of the Flow of an Anomalous Thermo-Viscous Fluid in an Annular Channel

A. A. Mukhutdinova, V. N. Kireev, and S. F. Urmancheev

Abstract In this paper, the flow of an anomalously thermo - viscous incompressible fluid in an annular channel, on the inner and outer surfaces of which convective heat transfer to the environment according to the Newton-Richman law, is considered. The viscosity of the fluid depends on the temperature in a non-monotone way. The mathematical model includes the continuity equation, Navier-Stokes equation, and temperature equation written in a cylindrical coordinate system with axial symmetry. As a result of numerical modeling, velocity diagrams in different cross sections of the annular channel and distributions of temperature and viscosity fields have been got. The influence of heat exchange conditions on the walls and rheological parameters of the fluid on the flow character are established. It is shown that a liquid with a non-monotone dependence of viscosity the hydrodynamic parameters of the flow depend on the location of the flow area with increased viscosity—the "viscous barrier".

12.1 Introduction

Possible changes in the thermophysical properties of liquids can have a significant impact on the flow regime and determine the functionality of technical systems. The dependence of viscosity on temperature, which is typical for most liquids, as a monotonically decreasing function, can be violated for substances in which the

A. A. Mukhutdinova (✉) · S. F. Urmancheev
Mavlyutov Institute of Mechanics, Ufa Investigation Center, R.A.S, Ufa, Russia
e-mail: mukhutdinova23@ya.ru

S. F. Urmancheev
e-mail: said52@mail.ru

V. N. Kireev
Ufa University of Science and Technology, Ufa, Russia
e-mail: kireevvn@uust.ru

H. Altenbach and V. Kuzkin (eds.), *Advanced Problems in Mechanics*,
https://doi.org/10.1007/978-3-031-80103-7_12

polymerization and depolymerization of molecules can occur in a limited temperature range. Such substances, for example, include liquid sulphur [1]. In industry, such substances with similar properties as liquid phosphorus, polymer solutions and melts, high sulphur and anomalously viscous oil are also widely used. Of particular interest is the work on the spreading of sulphur-bearing lava [2], which allowed the authors to draw important conclusions related to extraterrestrial volcanism. Detailed studies of various aspects of the flow of fluids with a nonmonotonic dependence of viscosity on temperature were undertaken in [3]. In connection with numerous possible applications, theoretical studies of hydrodynamic and heat transfer processes during the flow of anomalously thermo—viscous liquids seem to be unquestionably in demand.

12.2 Problem Statement

Consider the flow of a viscous incompressible fluid in an annular channel of length L formed by two coaxial circular cylinders (Fig. 12.1). The radius of the inner cylinder is r_0, the radius of the outer cylinder is R > r_0. Fluid flow occurs under the action of constant pressure drop Δp. The liquid at the inlet of the channel has a constant temperature T_0, and on the inner and outer walls of the channel there is heat exchange with the environment having constant temperature $T\infty < T_0$. Let us introduce a cylindrical coordinate system with z-axis directed along the axis of the cylinders.

The mathematical model consisting of the continuity equation, Navier-Stokes equations and the equation for temperature, written in cylindrical coordinates with axial symmetry in dimensionless form [4], has the following form:

$$\frac{\partial v_r}{\partial r} + \frac{\partial v_z}{\partial z} + \frac{v_r}{r} = 0 \tag{12.1}$$

$$\begin{gathered}\frac{\partial v_r}{\partial t} + v_r\frac{\partial v_r}{\partial r} + v_z\frac{\partial v_r}{\partial z} = \frac{\partial p}{\partial r} + \\ + \frac{1}{\mathrm{Re}}\left(\frac{\partial}{\partial r}\left(\mu(T)\frac{\partial v_r}{\partial r}\right) + \frac{\partial}{\partial z}\left(\mu(T)\frac{\partial v_r}{\partial z}\right) + \frac{\mu(T)}{r}\frac{\partial v_r}{\partial r} + \frac{\mu(T)}{r^2}v_r\right)\end{gathered} \tag{12.2}$$

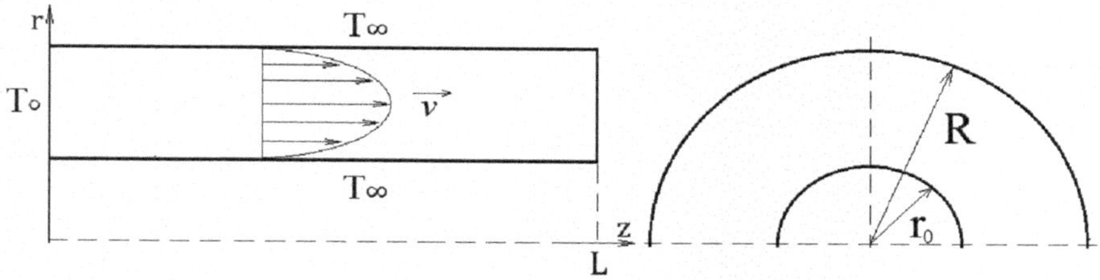

Fig. 12.1 Channel scheme

$$\frac{\partial v_z}{\partial t} + v_r \frac{\partial v_z}{\partial r} + v_z \frac{\partial v_z}{\partial z} = \frac{\partial p}{\partial z} + \\ + \frac{1}{\mathrm{Re}} \left(\frac{\partial}{\partial r} \left(\mu(T) \frac{\partial v_z}{\partial r} \right) + \frac{\partial}{\partial z} \left(\mu(T) \frac{\partial v_z}{\partial z} \right) + \frac{\mu(T)}{r} \frac{\partial v_z}{\partial r} \right) \tag{12.3}$$

$$\frac{\partial T}{\partial t} + v_r \frac{\partial T}{\partial r} + v_z \frac{\partial T}{\partial z} = \frac{1}{Pe} \left(\frac{\partial^2 T}{\partial r^2} + \frac{\partial^2 T}{\partial z^2} + \frac{1}{r} \frac{\partial T}{\partial r} \right) \tag{12.4}$$

where v_r and v_z are radial and axial components of the velocity vector, p is pressure, T is temperature, Re and Pe are dimensionless Reynolds and Peckle numbers.

The function $\mu(T)$ included in equations (12.2) and (12.3) is a temperature dependence of viscosity of liquid. The anomalous viscosity dependence on temperature (i.e. liquid viscosity depends on temperature in a non-monotonic way) [3] of the following form is considered in the present paper:

$$\mu(T) = \exp\left[-B(T - 0.5)^2\right] \tag{12.5}$$

where $B > 0$ is the parameter describing the nature of viscosity change.

Figure 12.2 shows viscosity dependence on temperature at different parameters B. Viscosity changes less as B decreases. The case with B equal to zero corresponds to constant viscosity.

For velocities on the channel walls, non-slip conditions are set. For temperature, convective heat transfer to the environment is set according to Newton—Richmann's law (third kind boundary condition):

$$\left.\frac{\partial T}{\partial r}\right|_{r=r_0/R} = \left.\frac{\partial T}{\partial r}\right|_{r=1} = -Nu \cdot T \tag{12.6}$$

where Nu is the dimensionless Nusselt number.

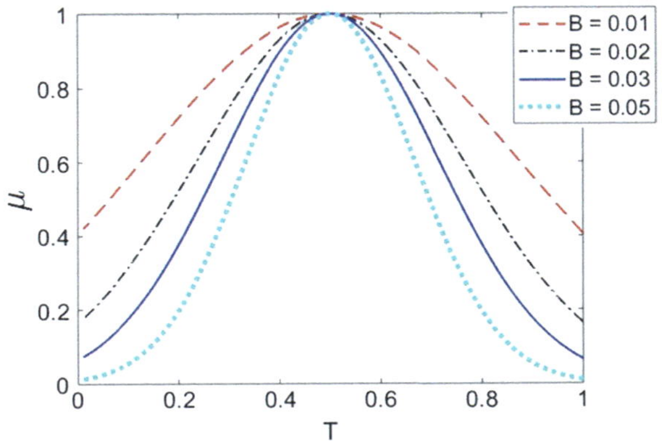

Fig. 12.2 Dependence of viscosity on temperature

12.3 Numerical Results

The mathematical model equations were solved numerically using the finite volume method and the SIMPLE (Semi-Implicit Method for Pressure-Linked Equation) algorithm [5] modified to take into account the variable viscosity coefficient. The original computer code has been implemented in C++ programming language in cross-platform development environment Qt Creator.

To validate the computer code, we consider the fluid flow with constant viscosity. In this case, we know the analytical solution to the problem [6]. Figure 12.3 shows the results of a comparison between the analytical and numerical solution. It can be seen that the numerical solution coincides with the analytical solution, the maximum relative error doesn't exceed two percent.

An anomalously viscous fluid flows in an annular channel of length L and height h under the action of applied differential pressure. We consider the channel to be highly elongated. The calculations, the results of which are given below, were carried out with the following values of dimensionless parameters:

$$L = 100R, \frac{h}{R} = \frac{1}{4}, B = 0.05, \mathrm{Re} = 948, Pe = 8540, Nu = 3, 10, 30, 200.$$

Figure 12.4 shows the distributions of temperature, viscosity, axial and radial velocity components of a steady flow in the axial section of the annular channel. It can be seen that the temperature field has a distribution typical for problems of convective heat conduction, which is due to the constant values of the heat capacity and heat conduction of the liquid under consideration. In accordance with the accepted nonmonotonic dependence of viscosity on temperature (5) and boundary conditions, the variable viscosity field will take the form shown in Fig. 12.4b, in which it is possible to distinguish a highly viscous region in the form of an arch—a viscous barrier, which will determine the features of the flow anomalously thermo—viscous

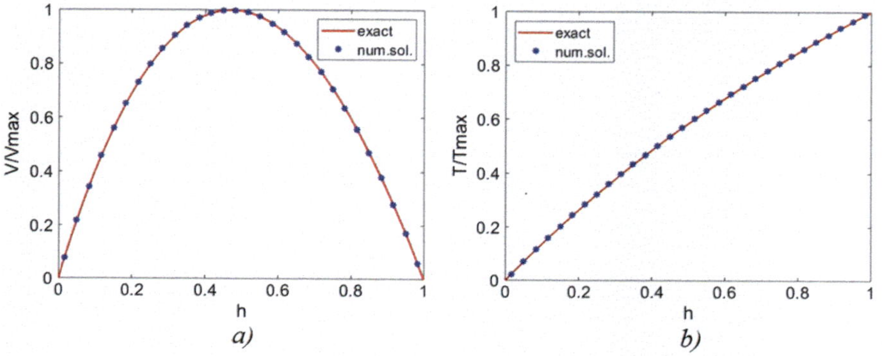

Fig. 12.3 Comparison between analytical and numerical solution for axial velocity distribution (**a**) and temperature (**b**)

liquid. The formation of a viscous barrier occurs in the process of interaction of a fluid flow with a temperature field, which ends with the establishment of a stable formation localized in space. In this calculation, the pressure drop was chosen so that the "viscous barrier" would fit entirely within the channel.

The inhomogeneity of the distribution of the physical parameters of the liquid leads to the formation of a jet in the flow and, accordingly, higher values of the longitudinal velocity in the central part of the axial section of the annular channel, located in front of the viscous barrier (Fig. 12.4c). The formation of a jet in the flow is also indicated by the characteristic distribution of the transverse component of the velocity field (Fig. 12.4d). Beyond the viscous barrier, the transverse velocity components disappear, and the longitudinal ones eventually form à Poiseuille distribution in the channel cross section.

Initially, it was natural to assume that the intensity of heat transfer would affect the distribution of all hydrodynamic parameters. Specific results of the study of the distribution of viscosity and pressure along the longitudinal coordinate at $r = (R - r_0)/2$ for various values of the Nusselt number are shown in Fig. 12.5. As seen from Fig. 12.5a, with a rather weak heat transfer ($Nu = 3$), the viscous barrier covers a

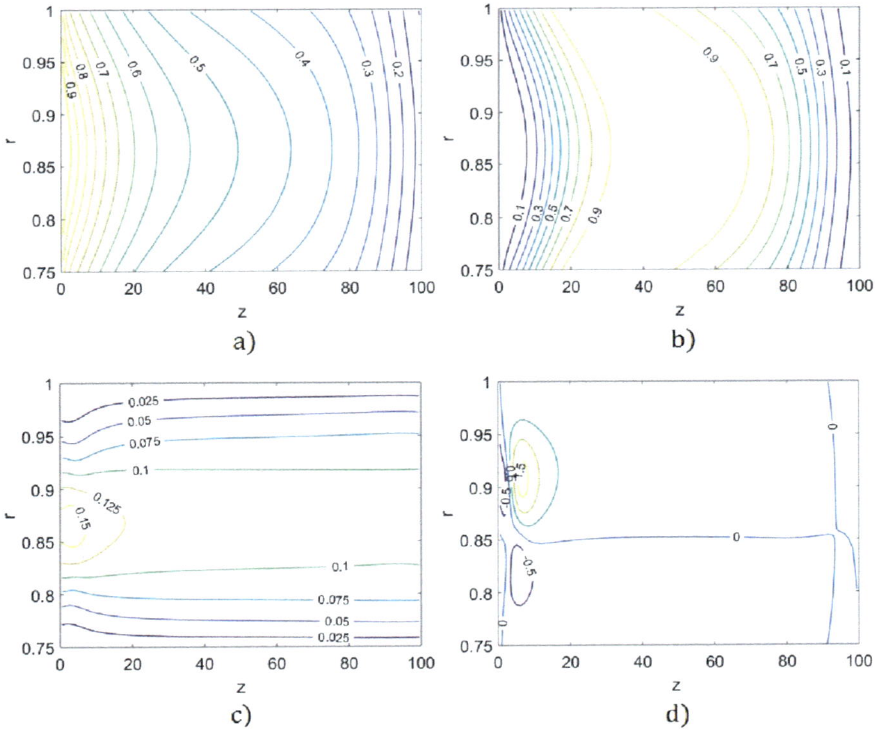

Fig. 12.4 The distribution of temperature **a**, viscosity **b**, axial **c** and radial **d** (10^{-4}) velocities along the axis of the annular channel at $Nu = 3$

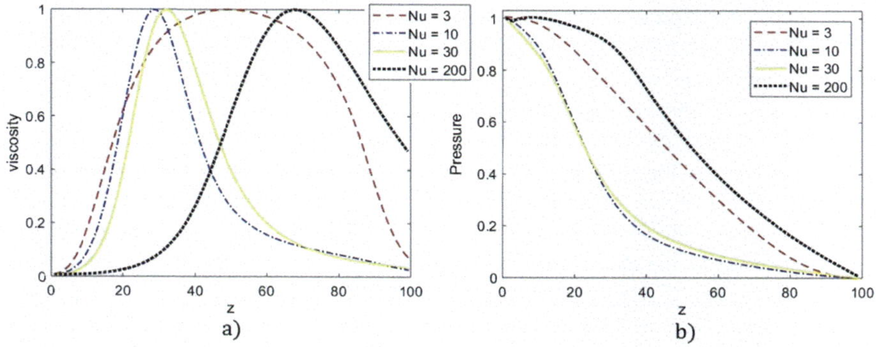

Fig. 12.5 The distribution of viscosity **a** and pressure **b** along the channel at different Nusselt numbers

significant part of the channel length. At intermediate values of the Nusselt number from 10 to 30 and, presumably, somewhat higher, the viscous barrier is concentrated in a narrower zone closer to the inlet section.

Further, with an increase in the intensity of heat transfer, the high-viscosity zone will move farther and farther from the inlet section. The extreme values correspond to either too slow or too fast cooling of the incoming liquid. We emphasize that the results of calculations for $Nu = 200$ are close to the calculation for boundary conditions of the first kind. The change in pressure is shown in Fig. 12.5b. Attention is drawn to the fact that in the calculations performed, the largest pressure drop is observed at Nusselt numbers equal to 10 and 30, in full accordance with the viscosity change profile.

From practical point of view the change of the flow rate is of great interest. On Fig. 12.6 the temporal variations of the flow rate for various values of the heat transfer intensity are shown. An increase in the final (steady-state) value of this parameter is noted with an increase in the Nusselt number.

In this case, not only the final value, but also the nature of the evolution of the liquid flow rate also depends on the intensity of heat transfer. At high values of the Nusselt numbers, the flow rate increases monotonically, practically reaching its maximum value in 200 s, then changing only by a small amount. In other cases, the evolution of the flow rate is nonmonotonic, due to the dynamics of the viscous barrier as a result of the interaction of hydrodynamic and thermal parameters.

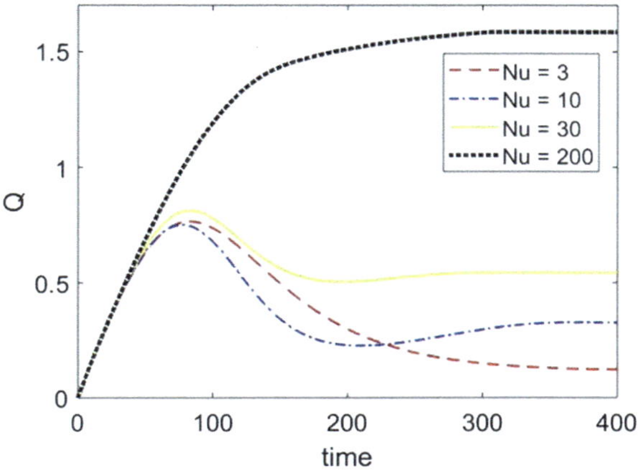

Fig. 12.6 Change in the flow rate over time for different Nusselt numbers

12.4 Conclusion

The obvious fact of the influence of the dependence of viscosity on temperature on the hydrodynamic parameters of a liquid and heat transfer is well studied and widely introduced into calculation practice. In this case, we are usually talking about monotonic dependencies that obey the Arrhenius law or its consequences. On the other hand, in the case of studying the influence of nonmonotonic dependences of the viscosity of a liquid on temperature on the nature of the flow, new regularities arise related to the interaction of heat fluxes with hydrodynamic fields, which are the subject of study in this work and impose on the flow parameters. The main conclusion is that the flow of an anomalously thermo—viscous liquid in a channel with an inhomogeneous temperature distribution is accompanied by the formation of a viscous barrier—a highly viscous region, in the cross section having the shape of an arch elongated along the flow.

The study was supported by the Russian Science Foundation (Project No. 22- 21-00915).

References

1. Schmidt M, Siebert W, Bugnall KW (1975) The chemistry of Sulphur, Selenium, Tellurium and Polonium. Pergamon texts in inorganic chemistry. Pergamon Press Oxford 15:795–1008
2. Fink JH, Park S, Greeley R (1983) Cooling and deformation of sulfur flows. Icarus 56, pp 38–50
3. Urmancheev SF, Kireev VN (2004) Steady flow of a liquid with a temperature anomalyof viscosity. Dokl Phys 49:328–331
4. Bird RB, Armstrong RC, Hassager O (1987) Dynamics of polymeric fluids. FluidMechanics 1:649

5. Patankar S (1980) Numerical heat transfer and fluid flow. Hemisphere Publishing Corporation, New York, pp 200
6. Slezkin NA (1955) Dynamics of a viscous incompressible fluid. Gostekhizdat, Moscow, pp 520 [in Russian]

Chapter 13
On the Use of Elastic Limiters in Two-Mass Vibrating Machines with Self-Synchronizing Inertial Vibration Exciters

Grigory Panovko and Alexander Shokhin

Abstract The dynamics of the technological vibrating machine model, designed according to the scheme of a two-mass chain oscillatory system, is considered. To excite vibrations, two self-synchronizing inertial vibration exciters fixed on the working element are used. Self-synchronization of vibration exciters and oscillations of a vibromachine near its natural frequencies with a slowly changing mass of the processed material on the working element are investigated. To expand the frequency range of the oscillatory system stability, an elastic element (an elastic vibration limiter) is used, installed with a clearance between the working element and the second movable mass. The interaction of the working element with the processed material is taken into account by an additional mass and a dissipative element. On the basis of numerical simulation, it has been established that the introduction of an elastic limiter leads to a significant expansion of the frequency range near the second resonant frequency, in which the synchronous antiphase rotation of vibration exciters required for the normal operation of the machine is realized. It has been established that due to the expansion of the operating frequency range, the working element vibrations excitation with higher amplitudes is provided. In addition, in this frequency range, the oscillation amplitudes are practically independent of the processed material mass, and their magnitude can be adjusted by the initial clearance between the limiter and the working element.

13.1 Introduction

There are many works devoted to the design and analysis of the dynamics of resonant-type vibromachines designed on the basis of two-mass dynamic schemes, for example [1–7]. The use of a two-mass dynamic scheme, in comparison with a single-mass

G. Panovko · A. Shokhin (✉)
Mechanical Engineering Research Institute of the Russian Academy of Sciences, Moscow, Russia
e-mail: shohinsn@mail.ru

H. Altenbach and V. Kuzkin (eds.), *Advanced Problems in Mechanics*,
https://doi.org/10.1007/978-3-031-80103-7_13

scheme, can significantly reduce the dynamic forces transmitted to the foundation while reducing the sensitivity of the working element oscillation amplitudes to changes in the characteristics of the processed material. To excite vibrations of resonant vibromachine, one vibration exciter is usually used, or several vibration exciters operating in the forced synchronization mode, for example, due to additional kinematic links or control circuits, due to which the required configuration of the exciting force is achieved [1, 3, 8].

The use of self-synchronizing vibration exciters can significantly simplify the design of the vibromachine and increase its reliability [2, 4]. At the same time, to create vibration excitation with a given configuration, it is necessary to adjust the system parameters depending on the choice of operating modes. A large number of works [9–15] are devoted to these issues, however, as a rule, modes of oscillations excitation far from resonant frequencies are considered.

The problem of using vibrating machines with self-synchronizing inertial vibration exciters in resonant modes is due to the high sensitivity of the mutual phase of their rotation and the working element vibrations amplitudes to changes in the processed material characteristics, which leads to a change in the exciting force configuration and to extremely undesirable jumps into the over-resonant frequency range [4, 10, 16, 17]. In some cases, these problems are solved by means of feedback control systems that correct the vibration exciters rotation frequency and automatically maintain the resonant tuning of the vibromachine in case of possible changes in the physical and mechanical characteristics of the processed material [16]. For vibromachines made according to a single-mass dynamic scheme, the possibility of stabilizing the synchronous-antiphase rotation of inertial vibration exciters and oscillations of the working element in the resonant frequency range due to the introduction of additional nonlinear elastic elements is shown in [18]. For the same purpose, in this paper, we consider the possibility of using additional nonlinear elastic elements for vibromachines designed according to a two-mass dynamic scheme.

13.2 Design Scheme and Mathematical Model of Vibromachine

Let us consider the design scheme of a two-mass vibrating machine with two inertial vibration exciters (Fig. 13.1). The oscillatory system of the machine consists of a working element with two inertial vibration exciters rigidly fixed on it, modeled by a solid body of mass m_2, and an intermediate platform, modeled by a solid body of mass m_1. A processed material with mass mw is located on the working element. The platform is connected to the fixed base by two identical elastic elements (springs) with linear characteristics of stiffness $c_1/2$ and viscous friction $b_1/2$. The working element and the platform are interconnected by two identical springs with linear characteristics of stiffness $c_2/2$ and viscous friction $b_2/2$. A linearly elastic vibration limiter with a stiffness coefficient c_l is installed on the working element towards the

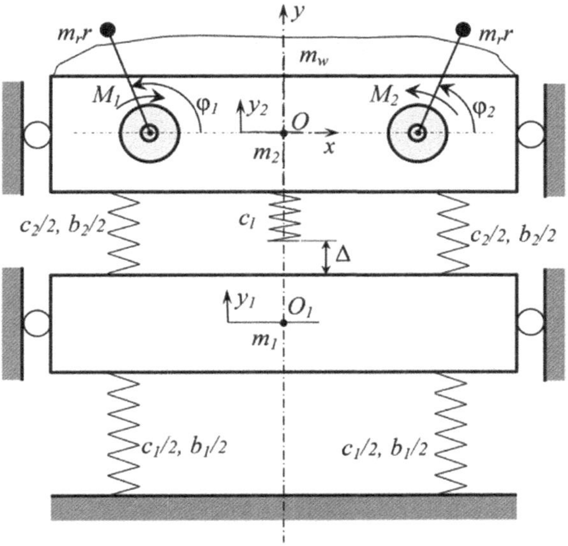

Fig. 13.1 Design scheme

platform, which "turns on" when the movement of the working element relative to the platform exceeds the clearance Δ. The working element and the platform can only move in the vertical direction. Gravity is not taken into account.

Vibrations of the system are excited by two inertial vibration exciters driven by asynchronous type electric motors rotating in opposite directions. Vibration exciters have the same static mass moments $m_r r$ and practically the same characteristics of driving torque $M_j(s)$, depending on the slip $s(\dot{\varphi}_j)$ of the induction motor ($\dot{\varphi}_j$—is the rotation speed of the j-th vibration exciter), which are described by the Kloss formula [16]. Thus, with synchronous antiphase rotation of vibration exciters, a unidirectional exciting force arises, under the action of which the system will perform unidirectional oscillations.

The interaction of the processed material with the working element is taken into account by the friction force $F_w(n)$, which expresses the dissipation of energy in the processed material, where $n = k_T m_w / \gamma_M$—is the normalized dissipation coefficient, k_T—is the specific (per unit volume) coefficient of viscous friction, γ_M—is the material bulk density [2, 4].

The movement of the system is described by the displacements of the mass centers of the platform $y_1(t)$ and the working element $y_2(t)$ relative to their undeformed position, and the angles of rotation of the vibration exciters disbalances $\varphi_j(t)$, counted in a counterclockwise direction from the positive direction of the Ox axis.

Taking into account the elastic limiter, the differential equations of motion are written in the following form:

$$\begin{cases} m_1\ddot{y}_1 + (b_1 + b_2)\dot{y}_1 - b_2\dot{y}_2 + (c_1 + c_2)y_1 - c_2y_2 - F_l = 0, \\ (m + m_w)\ddot{y}_2 - b_2\dot{y}_1 + b_2\dot{y}_2 - c_2y_1 + c_2y_2 + F_l + F_w = m_r r \sum_{j=1}^{2} \left(\dot{\varphi}_j^2 \sin\varphi_j - \ddot{\varphi}_j \cos\varphi_j\right), \\ J\ddot{\varphi}_j = M_j(\dot{\varphi}_j) - m_r r\ddot{y}_2 \cos\varphi_j, \end{cases}$$

where $F_l = \begin{cases} c_l(y_2 - y_1 - \Delta), y_2 - y_1 \leq \Delta \\ 0, y_2 - y_1 > \Delta \end{cases}$—the elastic limiter reaction force, $m = m_2 + 2m_r, J = J_r + m_r r^2, J_r$—moment of inertia of the vibration exciter shaft relative to the axis of rotation, $M_j(\dot{\varphi}_j) = 2\sigma_j M_{cj}/(s_{cj}/s_j + s_j/s_{cj})$—electric motors driving torques, $j = 1, 2$—the vibration exciter index, M_{cj} and s_{cj} respectively, the critical torque and slip of the j-th electric motor, $s_j = (\omega_{0j} - \sigma_j\dot{\varphi}_j)/\omega_{0j}$, $\omega_{0j} = 2\pi f_e/P$—synchronous motor speed, f_e – power supply frequency, P—number of motor pole pairs, $\sigma_j = \pm 1$—coefficient taking into account the direction of the torque ($\sigma_j = +1$ corresponds to counterclockwise action of the motor torque), $F_w = n\dot{y}_2$—dissipative force arising in the processed material volume under vibration.

13.3 Numerical Simulation Results

The analysis of the influence of the elastic limiter and its parameters on the system dynamics was performed for various values of the processed material mass $m_w = [0, 0.4]m$ in the range of electric motor power supply frequencies $f_e = [2, 80]$ Hz at various values of the clearance Δ. Integration of the equations of motion of the system was performed numerically by the Runge–Kutta method with determination of the time points of occurrence and loss of contact with the elastic limiter (condition $y_2 - y_1 = \Delta$) necessary for correctly taking into account the reaction force of the limiter F_l. The selected range of change in the material mass is determined by the restrictions on the maximum load on the machine working element, and the range of power supply frequency change covers the range of all resonant frequencies for the selected machine parameters: $m = m_l = 12.5$ kg, $m_r r = 6.7 \times 10^{-3}$ kg·m, $J = 1.2$ 10^{-3} kg m^2, $c_2 = 210.2$ kN/m, $c_l = 0.1c_2$, $b_l = b_2 = 97.25$ N s/m, $M_{crl} = M_{cr2} =$ 1.9 Nm, $s_{crl} = s_{cr2} = 0.15$, $P = 2$, k_T/γ_M=82 s^{-1}, $c_l = 8c_2$.

The results are presented in the form of graphs of the working element oscillations extrema $y_{2,max}$ and $y_{2,min}$, and the mutual phase of the vibration exciters rotation $\Delta\varphi$ depending on their synchronous frequency of rotation f for various values of m_w, built in the case of absence of a limiter (Figs. 13.2 and 13.3), and in the case of contact with the limiter (Figs. 13.4 and 13.5).

The oscillatory system under consideration has two significant degrees of freedom, and accordingly has two eigenfrequencies and eigenmodes, the first of which (the smallest one) corresponds to the in-phase motion of solid bodies, and the second one to their antiphase motion. When the frequency of the disturbing force coincides with one of the eigenfrequencies, resonant oscillations of the system occur, characterized by a significant increase in their amplitudes. With the selected system parameters

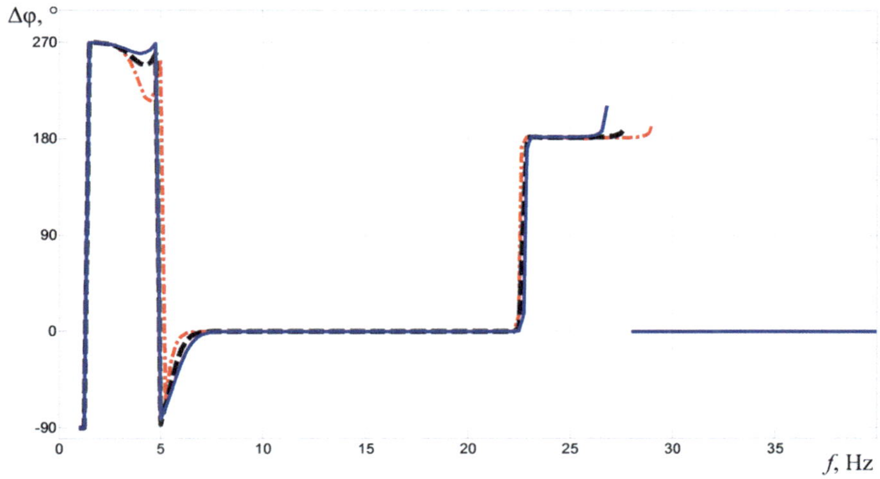

Fig. 13.2 Mutual phase shift of vibration exciters rotation depending on the excitation frequency in the case of absence of a limiter: the red dashed-dotted line corresponds to the mass of the material $m_w = 0$, the black dashed line—$m_w = 0.2$ m, the blue solid line—$m_w = 0.4$ m

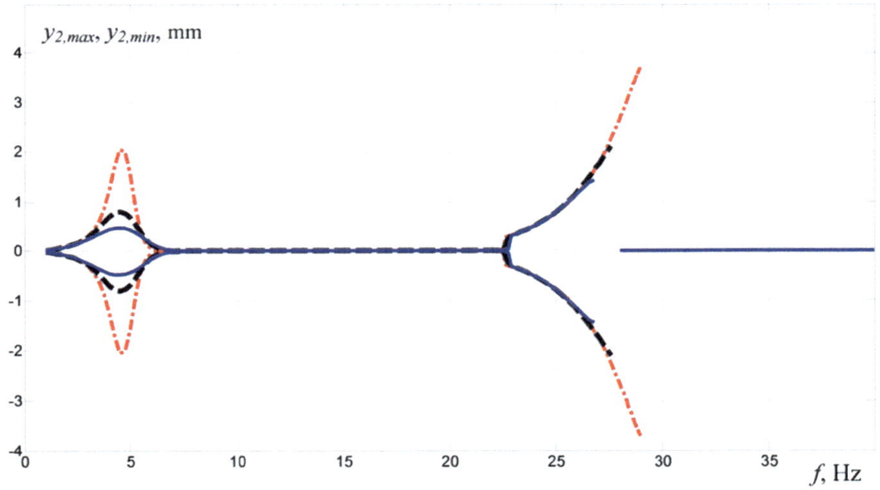

Fig. 13.3 Working element oscillations extrema depending on the excitation frequency in the case of absence of a limiter: the red dashed-dotted line corresponds to the mass of the material $m_w = 0$, the black dashed line—$m_w = 0.2$ m, the blue solid line—$m_w = 0.4$ m

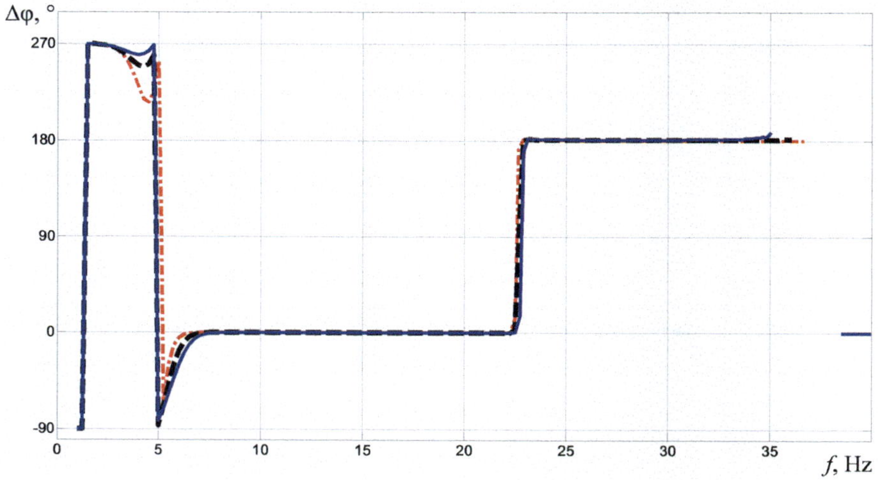

Fig. 13.4 Mutual phase shift of vibration exciters rotation depending on the excitation frequency in the case of contact with the limiter installed with a clearance of $\Delta = 2$ mm: the red dashed-dotted line corresponds to the mass of the material $m_w = 0$, the black dashed line—$m_w = 0.2$ m, the blue solid line—$m_w = 0.4$ m

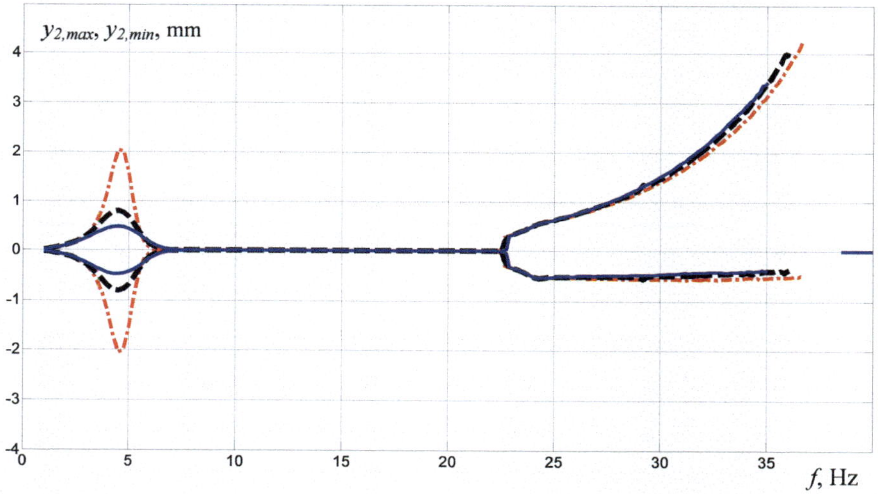

Fig. 13.5 Working element oscillations extrema depending on the excitation frequency in the case of contact with the limiter installed with a clearance of $\Delta = 2$ mm: the red dashed-dotted line corresponds to the mass of the material $m_w = 0$, the black dashed line—$m_w = 0.2$ m, the blue solid line—$m_w = 0.4$ m

and in the absence of the processed material ($m_w = 0$), the first resonance occurs at a frequency of 4.5 Hz, and the second – at a frequency of 29 Hz. In this case, a stable synchronous-antiphase rotation of vibration exciters with mutual phase $\Delta\varphi \approx 180°$ is realized in the frequency range between the first and second resonances of the system (red dash-dotted line in Fig. 13.2). Near the first resonant frequency, due to small oscillation amplitudes (Fig. 13.3), the synchronization conditions are practically not fulfilled. An increase in the material mass leads to a noticeable deviation of the mutual phase from 180° near the second resonance (dashed black and solid blue lines in Fig. 13.2), which is due to an increase in dissipative forces in the processed material volume due to an increase in vibration amplitudes. In addition, the second resonant frequency of the system decreases, which leads to a narrowing of the operating frequency range and an earlier jump into the over-resonant mode. In contrast to the frequency range near the first resonant frequency of the system, when oscillations are excited near the second resonant frequency, the amplitudes of the working element oscillations weakly depend on the material mass. Thus, the use of resonant modes near the second resonant frequency as working ones is mainly difficult due to the risk of oscillation jump to the over-resonant frequency range. This leads to a significant narrowing of the permissible range of operating frequencies, in which relatively small amplitudes of the working element oscillations can be realized.

Figures 13.4 and 13.5 show the graphs of mutual phase $\Delta\varphi$ and extrema of the working element vibrations $y_{2,max}$ and $y_{2,min}$ depending on the frequency f in the case of contact with the limiter installed with a clearance of $\Delta = 2$ mm. The clearance value is chosen so that contact with the limiter is guaranteed at the maximum allowable mass of the material.

The introduction of an elastic limiter leads to a significant expansion of the frequency range near the second resonant frequency, in which a stable synchronous-antiphase rotation of vibration exciters with mutual phase $\Delta\varphi \approx 180°$ (Fig. 13.4) is realized due to the shift of its upper boundary to the right. The oscillation amplitudes (Fig. 13.5) and the mutual phase of rotation of the vibration exciters (Fig. 13.4) in this frequency range turn out to be practically insensitive to changes in the material mass. In this case, near the upper boundary of this frequency range, significantly higher oscillation amplitudes can be achieved than in the absence of a limiter, which is associated with an increase in the exciting force developed by inertial vibration exciters with an increase in rotational speed. An increase in the initial clearance Δ leads to a narrowing of this frequency range and a decrease in the possible amplitudes of the working element oscillations.

It should be noted that issues related to the use of nonlinear elastic elements to ensure constancy of the resonant frequency and oscillations amplitudes when the system mass changes were discussed, for example, in [19], however, the use of such elements is not sufficiently effective for oscillations with relatively large amplitudes that arise in the problem under consideration.

13.4 Conclusion

The performed analysis showed that the introduction of an elastic limiter leads to the expansion of the frequency range near the second resonant frequency of the system, in which the stable synchronous-antiphase rotation of vibration exciters required for the normal operation of the vibromachine is realized. In this frequency range, there is a weak sensitivity of the shape and amplitudes of working element vibrations to changes in the processed material mass. In addition, the introduction of an elastic limiter also leads to the possibility of excitation higher amplitudes of the working element oscillations. The value of the initial clearance between the limiter and the working element has a significant impact on both the oscillations amplitude and the boundaries of the frequency range of stable synchronous-antiphase rotation of vibration exciters, and can be used to adjust the oscillations amplitude at a given excitation frequency. Thus, these near-resonant oscillatory modes, which are characterized by low sensitivity to changes in the material mass, can be used as working ones without the use of additional systems for controlling the frequency and amplitude of excitation.

Acknowledgements This research is funded by the Russian Science Foundation, project No. 21-19-00183, https://rscf.ru/en/project/21-19-00183/.

References

1. Poturaev VN, Chervonenko AG, Obodan YY (1989) Dynamics and strength of transport-technological vibration machines. Mashinostroenie, Leningrad
2. Blekhman II (2013) Theory of vibration processes and devices. Vibration mechanics and vibration technology. Ore and Metals PH, St. Petersburg
3. Handbook: vibration in technics (1981) Vibration processes and machines. In: Levendel EE (ed) Mashinostroenie, Moscow, vol 4
4. Vaisberg LA (1986) Design and calculation of vibrating screens. Nedra, Moscow
5. Tsyfanskii SL, Oks AB, Beresnevich VI (1991) Nonlinear and parametric vibrations of technological vibration machines. Zinatne, Riga
6. Dentsov NN, Koshelev AV (2018) Dynamics of a dual-mass resonant vibration screen in a first approximation. Vestnik of Samara University. Aerosp Mech Eng 17(3):148–157
7. Zhang XL, Kong XX, Wen BC, Zhao CY (2015) Numerical and experimental study on synchronization of two exciters in a nonlinear vibrating system with multiple resonant types. Nonlinear Dyn 82(1–2):987–999
8. Huang Z, Song G, Zhang Z, Zhang X (2019) Control synchronization of two nonidentical homodromy exciters in nonlinear coupled vibration system. IEEE Access 7:109934–109944
9. Blekhman II (1971) Synchronization of dynamical systems. Nauka, Moscow
10. Zhang X, Wen B, Zhao C (2016) Theoretical study on synchronization of two exciters in a nonlinear vibrating system with multiple resonant types. Nonlinear Dyn 85(1):141–154
11. Zhang N (2020) Self-synchronization characteristics of a class of nonlinear vibration system with asymmetrical hysteresis. J Low Freq Noise Vib Act Control 39(1):114–128
12. Nijmeijer H, Rodriguez-Angeles A (2003) Synchronization of mechanical systems. World Scientific Publishing, Singapore

13. Nagaev R (2003) Dynamics of synchronizing systems. Springer-Verlag, Berlin
14. Liu Y, Zhang X, Gu D, Jia L, Wen B (2020) Synchronization of a dual-mass vibrating system with two exciters. Shock and vibration. Art ID 9345652
15. Vasilkov VB, Shishkin YV (2014) Dynamics of a vibrating device with torsionally suspended pendulums. Obogashchenie Rud 6:25–28
16. Panovko GY, Shokhin AE (2020) Dynamics of resonant vibration machines with self-synchronizing unbalanced vibration exciters. Institute of computer sciences, Moscow-Izhevsk
17. Yong-Zheng J, Kuan-Fang H, Yong-Le D, Da-Lian Y, Wei S (2019) Influence of load weight on dynamic response of vibrating screen. Shock and vibration. Art ID 4232730
18. Panovko G, Shokhin A (2021) Dynamics features of a resonant vibromachine with self-synchronizing inertial vibroexciters in the presence of vibration limiter. J Vibroengineering 23(8):1727–1736
19. Altshul GM, Guskov AM, Panovko GY (2022) Dynamics of a resonant vibrator with an equal-frequency suspension of the working body and an unbalanced vibration exciter. Obogashchenie Rud 1:51–55

Chapter 14
Applied Problems of Static Transvers Bending, Stability and Vibrations of a Graphene Sheet

S. H. Sargsyan

Abstract In this work, on the basis of the previously developed continuous moment-membrane theory for describing deformations of two-dimensional nanomaterials (including graphene), three applied problems are considered for a rectangular graphene sheet: (1) the problem of static transverse bending, (2) the problem of stability of its initial compressed state, (3) the problem of its natural transverse bending vibrations. For the problem of static transverse bending of a graphene sheet, the functions of displacement and free rotations, internal force and moment characteristics are determined. In the case of the problem of the stability of the initially compressed state of a graphene sheet, the critical load is determined. When studying the problem of its natural vibrations, the frequencies of natural transverse-bending vibrations are determined, showing that the lowest natural vibration frequency of a graphene sheet is in giga- terahertz (GHz, THz) frequency region.

14.1 Introduction

Methods of mechanics of a deformable solid body are widely used to model deformations of a nanostructure. The construction of continuous models of deformations of two-dimensional nanomaterials, in particular that of graphene, is one of the topical problems in the mechanics of deformable solids. It is established [1] that the deformation of crystalline nanomaterials takes place based on the "shear plus rotation" deformation model. In [2–4] it is also established that for the study of deformations of two-dimensional nanomaterials according to the continuous theory, the use of the three-dimensional moment theory of elasticity with independent fields of displacements and rotations is appropriate.

The above-mentioned studies facilitate the development of the continuous theory of deformations of two-dimensional nanomaterials as an adequate theory of plates

S. H. Sargsyan (✉)
State University of Shirak, Gyumri, Armenia
e-mail: s_sargsyan@yahoo.com

H. Altenbach and V. Kuzkin (eds.), *Advanced Problems in Mechanics*,
https://doi.org/10.1007/978-3-031-80103-7_14

and shells according to the moment theory of elasticity. In [5], a continuous one-dimensional beam model of a linear atomic chain is constructed, when the atomic interaction in its discrete model are of force non-central and moment character. With the application of the mentioned model, the paper further demonstrates the construction of the discrete–continuous model, replacing the interaction between the atoms of a two-dimensional nanomaterial with a beam system. On the example of a graphene, by passing to the limit, two continuous models of deformation are constructed: (1) a model of a plane stress state of a graphene sheet, (2) a model of its transverse bending, with the establishment that both of these models are identical to the corresponding models of the moment-membrane theory of elastic plates [6, 7] and all the six elastic constants of the moment theory of elasticity for the graphene material are determined. Thus, we can state that the moment-membrane models of a plane stress state and the transverse bending of elastic plates, with already known elastic constants, can be interpreted as continuous models for the corresponding deformations of a graphene sheet.

It is evident that the mentioned continuous models open up new avenues for setting and studying various applied problems for graphene sheet deformations. In this paper, for the transverse bending of a graphene sheet, the following applied problems are studied: (1) the problem of static transverse bending of a rectangular graphene sheet, with transverse load distribution, and freely supported edges of the sheet, where we determine the bending of the graphene sheet, as well the acting forces and moments; (2) the problem of stability for the initially compressed state of the graphene sheet, where we determine the critical load; (3) the problem of the free transverse bending vibrations of a graphene sheet, where we determine the frequencies of natural vibrations and show that the minimum natural vibration frequency of a graphene sheet is in approximate giga- terahertz (GHz, THz) frequency region.

14.2 Basic Equations and Boundary Conditions of the Moment-Membrane Theory of Elastic Thin Plates

To study the deformations of a graphene sheet according to the continuous theory, from the general equations and relations of the moment-membrane theory of elastic thin shells [6, 7], we move on to a particular theory–the theory of plates. As a result, two separate theories are obtained for plates: (1) a model of a plane stress state of the moment-membrane theory of elastic thin plates, (2) a model of transverse bending of the moment-membrane theory of elastic thin plates. Using these models, we can separately study the deformation of a graphene sheet in its own plane and bending deformation from its plane.

Our immediate goal is to study these models.

1) Model of the plane stress state of the moment-membrane theory of elastic thin plates.

The basic equations of the plane stress state of the moment-membrane theory of elastic thin plates can be obtained from the general equations and relations of the theory of shells [6, 7]. In Cartesian coordinates (x, y), these equations are expressed as follows:

Equilibrium (motion) equations:

$$\begin{gathered}\frac{\partial T_{11}}{\partial x}+\frac{\partial S_{21}}{\partial y}=\rho_0\frac{\partial^2 u_1}{\partial t^2}-\left(p_1^+-p_1^-\right),\\ \frac{\partial S_{12}}{\partial x}+\frac{\partial T_{22}}{\partial y}=\rho_0\frac{\partial^2 u_2}{\partial t^2}-\left(p_2^+-p_2^-\right),\\ \frac{\partial L_{13}}{\partial x}+\frac{\partial L_{23}}{\partial y}+(S_{12}-S_{21})=I_0\frac{\partial^2 \Omega_3}{\partial t^2}-\left(m_3^+-m_3^-\right);\end{gathered}\tag{14.1}$$

Geometric relations:

$$\begin{gathered}\Gamma_{11}=\frac{\partial u_1}{\partial x},\ \Gamma_{22}=\frac{\partial u_2}{\partial y},\ \Gamma_{12}=\frac{\partial u_2}{\partial x}-\Omega_3,\\ \Gamma_{21}=\frac{\partial u_1}{\partial y}+\Omega_3,\ k_{13}=\frac{\partial \Omega_3}{\partial x},\ k_{23}=\frac{\partial \Omega_3}{\partial y};\end{gathered}\tag{14.2}$$

Physical relations of elasticity:

$$\begin{gathered}T_{11}=\frac{2Eh}{1-\nu^2}(\Gamma_{11}+\nu\Gamma_{22}),\ T_{22}=\frac{2Eh}{1-\nu^2}(\Gamma_{22}+\nu\Gamma_{11}),\\ S_{12}=(\mu+\alpha)\cdot 2h[\Gamma_{12}+\eta\Gamma_{21}],\\ S_{21}=(\mu+\alpha)\cdot 2h[\Gamma_{21}+\eta\Gamma_{12}],\\ L_{13}=2Bhk_{13},\ L_{23}=2Bhk_{23},\ \eta=\frac{\mu-\alpha}{\mu+\alpha},\ B=\frac{4\gamma\varepsilon}{\gamma+\varepsilon};\end{gathered}\tag{14.3}$$

Boundary conditions, for example on $y = const$, are expressed as follows:

a) Static boundary conditions:

$$T_{11}=\overline{T}_{11},\ S_{12}=\overline{S}_{12},\ L_{13}=\overline{L}_{13},\tag{14.4}$$

b) Geometric boundary conditions:

$$u_1=\overline{u}_1,\ u_2=\overline{u}_2,\ \psi_3=\overline{\psi}_3.\tag{14.5}$$

Mixed boundary conditions can also take place.

It should be noted that in [8] the specified model of plane deformation of a plate is known according to the moment theory of elasticity with independent fields of displacements and rotations.

Let us reduce the initial equations and relations (14.1)-(14.3) of the plane stress state model of the moment-membrane theory of elastic thin plates to resolving equations. Substituting (14.2) into (14.3) and then into the equilibrium Eqs. (14.1), we obtain a system of resolving equations for the functions $u_1(x,y)$, $u_2(x,y)$, $\Omega_3(x,y)$:

$$(\lambda'+2\mu)\frac{\partial}{\partial x}\left(\frac{\partial u_1}{\partial x}+\frac{\partial u_2}{\partial y}\right)+(\mu+\alpha)\left(\frac{\partial^2 u_1}{\partial y^2}-\frac{\partial^2 u_2}{\partial x\partial y}\right)+2\alpha\frac{\partial\Omega_3}{\partial y}=-p_1,$$

$$(\lambda'+2\mu)\frac{\partial}{\partial y}\left(\frac{\partial u_1}{\partial x}+\frac{\partial u_2}{\partial y}\right)+(\mu+\alpha)\left(\frac{\partial^2 u_2}{\partial x^2}-\frac{\partial^2 u_1}{\partial x\partial y}\right)-2\alpha\frac{\partial\Omega_3}{\partial x}=-p_2,$$

(14.6)

$$B\Delta\Omega_3-4\alpha\Omega_3+2\alpha\left(\frac{\partial u_2}{\partial x}-\frac{\partial u_1}{\partial y}\right)=-m_3,$$

where

$$\lambda'=\frac{2\lambda\mu}{\lambda+2\mu}=\frac{\nu}{1-\nu^2},\quad \Delta(\cdot)=\frac{\partial^2(\cdot)}{\partial x^2}+\frac{\partial^2(\cdot)}{\partial y^2}. \tag{14.7}$$

The system of Eqs. (14.6) is supplemented with boundary conditions (14.4) or (14.5).

2) Model of transverse bending of the moment-membrane theory of elastic thin plates.

The basic equations and boundary conditions of the transverse bending model of the moment-membrane theory of elastic thin plates, which follow from the general equations and boundary conditions of the theory of thin shells [6, 7], are introduced below in the Cartesian coordinate system (x, y), assuming that only a distributed load $p_3^+ - p_3^- = p_3$ normal to the median plane acts on the plate.

Equilibrium (motion) equations:

$$\frac{\partial N_{13}}{\partial x}+\frac{\partial N_{23}}{\partial y}=2\rho_0 h\frac{\partial^2 w}{\partial t^2}-p_3,$$

$$\frac{\partial L_{11}}{\partial x}+\frac{\partial L_{21}}{\partial y}+N_{23}=2I_0 h\frac{\partial^2\Omega_1}{\partial t^2}, \tag{14.8}$$

$$\frac{\partial L_{12}}{\partial x}+\frac{\partial L_{22}}{\partial y}-N_{13}=2I_0 h\frac{\partial^2\Omega_2}{\partial t^2};$$

Physical relations of elasticity:

$$N_{13} = 2G^*h\Gamma_{13},\ N_{23} = 2G^*h\Gamma_{23},\ G^* = \frac{4\mu\alpha}{\mu+\alpha}$$

$$L_{11} = 2h\frac{2\gamma}{\beta+2\gamma}\left[2(\beta+\gamma)k_{11} + \beta\cdot k_{22}\right],$$

$$L_{22} = 2h\frac{2\gamma}{\beta+2\gamma}\left[2(\beta+\gamma)k_{22} + \beta\cdot k_{11}\right], \tag{14.9}$$

$$L_{12} = 2h\left[(\gamma+\varepsilon)k_{12} + (\gamma-\varepsilon)k_{21}\right],$$

$$L_{21} = 2h\left[(\gamma+\varepsilon)k_{21} + (\gamma-\varepsilon)k_{12}\right];$$

Geometric relations:

$$\begin{gathered}\Gamma_{13} = \frac{\partial w}{\partial x} + \Omega_2,\ \Gamma_{23} = \frac{\partial w}{\partial y} - \Omega_1,\\ k_{11} = \frac{\partial\Omega_1}{\partial x},\ k_{22} = \frac{\partial\Omega_2}{\partial y},\ k_{12} = \frac{\partial\Omega_2}{\partial x},\ k_{21} = \frac{\partial\Omega_1}{\partial y};\end{gathered} \tag{14.10}$$

Static boundary conditions (for example on $y = const$):

$$N_{13} = \overline{N}_{13},\ L_{11} = \overline{L}_{11},\ L_{12} = \overline{L}_{12}, \tag{14.11}$$

Geometrical boundary conditions:

$$w = \overline{w},\ \Omega_1 = \overline{\Omega}_1,\ \Omega_2 = \overline{\Omega}_2 \tag{14.12}$$

Mixed boundary conditions also can take place.

Now we reduce the initial equations and relations (14.8)-(14.10) of the transverse bending model of the moment-membrane theory of elastic thin plates to resolving equations. Substituting (14.10) into (14.9) and then into the equilibrium Eqs. (14.8), we obtain a system of differential equations for the functions $w(x,y), \Omega_1(x,y), \Omega_2(x,y)$:

$$\begin{gathered}\Delta w + \left(\frac{\partial\Omega_2}{\partial x} - \frac{\partial\Omega_1}{\partial y}\right) = -\frac{p_3}{D_*} + \frac{\rho_0}{D_*}\frac{\partial^2 w}{\partial t^2},\\ \Delta\Omega_1 + \nu_m\frac{\partial}{\partial x}\left(\frac{\partial\Omega_1}{\partial x} + \frac{\partial\Omega_2}{\partial y}\right) + \frac{D_*}{D'}\left(\frac{\partial w}{\partial y} - \Omega_1\right) = \frac{I_0}{D'}\frac{\partial^2\Omega_1}{\partial t^2},\\ \Delta\Omega_2 + \nu_m\frac{\partial}{\partial y}\left(\frac{\partial\Omega_1}{\partial x} + \frac{\partial\Omega_2}{\partial y}\right) - \frac{D_*}{D'}\left(\frac{\partial w}{\partial x} + \Omega_2\right) = \frac{I_0}{D'}\frac{\partial^2\Omega_2}{\partial t^2},\end{gathered} \tag{14.13}$$

where

$$D_* = 2G^*h,\ D' = 2h(\gamma + \varepsilon),\ \nu_m = 2\frac{\gamma - \varepsilon}{\gamma + \varepsilon}. \tag{14.14}$$

Note that when obtaining the system of resolving equations for the bending model of the moment-membrane theory of elastic thin plates, the following formula [5] was taken into account:

$$\beta = \frac{\gamma - \varepsilon}{\gamma + \varepsilon} \cdot 2\gamma. \tag{14.15}$$

It is easy to see that the system of equations for the bending model of the moment-membrane theory of elastic thin plates, from a mathematical point of view, is identical with the system of equations for the bending deformation of elastic thin plates of Timoshenko type [9].

When there are complex boundary conditions (for example, when a plate, in this case, a sheet of graphene, is weakened by holes or there are fractures, cutouts, insertions), then it is convenient to reduce the system of Eqs. (14.13) of the bending model of the moment-membrane theory of elastic thin plates to a system of two equations–to the biharmonic problem and to the Helmholtz equation.

For this purpose, from the last two equations of system (14.13), we eliminate the combination $\left(\frac{\partial \Omega_2}{\partial x} - \frac{\partial \Omega_1}{\partial y}\right)$, for which we differentiate the second equation from (14.13) by y, and the third by x and, subtract the resulting ones:

$$(k\Delta - 1)\left(\frac{\partial \Omega_2}{\partial x} - \frac{\partial \Omega_1}{\partial y}\right) = \Delta w, \tag{14.16}$$

where

$$k = \frac{D'}{D_*}. \tag{14.17}$$

Using the first equation from system (14.13), as well as Eq. (14.16), with respect to the deflection $w(x, y)$, we arrive at the following equation:

$$\Delta\Delta w = \frac{p_3}{D'} - \frac{1}{D_*}\Delta p_3. \tag{14.18}$$

On the basis of the paper [9], functions ψ and ϕ are introduced with the help of formulas

$$\Omega_1 = -\frac{\partial \phi}{\partial y} + \frac{\partial \psi}{\partial x},\ \Omega_2 = \frac{\partial \phi}{\partial x} + \frac{\partial \psi}{\partial y}, \tag{14.19}$$

where the function ϕ is expressed through w and p_3:

$$\phi = -w - \frac{D'}{D_*}\Delta w - \frac{D'}{D_*^2}p_3. \tag{14.20}$$

The last two equations from (14.13) will be satisfied if the function ψ satisfies Helmholtz equation:

$$\Delta\psi - \tilde{k}^2\psi = 0, \tag{14.21}$$

where

$$\tilde{k}^2 = \frac{D_*}{(1+\nu_m)D'}. \tag{14.22}$$

Thus, the system of resolving equations for the model of transverse bending of the moment-membrane theory of elastic thin plates, in this case, bending of a graphene sheet from its plane, will have the form:

$$D'\Delta\Delta w = p_3 - k\Delta p_3, \; \Delta\psi - \tilde{k}^2\psi = 0. \tag{14.23}$$

Forces and moments will be expressed in functions w, ψ, with the consideration of formulas (14.9), (14.10), (14.19). It can be shown that if the plate is hinged along the entire outer contour, then the function $\psi \equiv 0$.

14.3 Elastic Constants of the Moment-Membrane Theory of Plane Stress State and Transverse Bending Deformation of Plates for Graphene Material

In [5], after determination of the potential energy of grapheme deformation, under a plane stress state and under transverse bending deformation, these expressions are compared with the corresponding expressions of the moment-membrane continuous theory of elastic plates. As a result, the elastic constants of the moment-membrane continuous theory of plates are determined (for the two specified deformations) through the physical parameters of graphene in its discrete model.

Thus, in the plane stress state of the moment-membrane theory of elastic plates, for the elastic constants of the graphene material we will have [5]:

$$E_* = 2Eh = 287\frac{N}{m}; \; \mu_* = 2\mu h = 116\frac{N}{m};$$

$$\alpha_* = 2\alpha h = 42\frac{N}{m}; \; \nu = 0,24; \tag{14.24}$$

$$B_* = 2Bh = 5,05 \cdot 10^{-10} N \cdot nm.$$

In the case of transverse bending deformation, the elastic constants of the moment-membrane theory of plates for the graphene material are [5]:

$$D_* = 2G_*h = 86\frac{N}{m},\ D' = 2h(\gamma+\varepsilon) = \gamma^* + \varepsilon^* = 4,15\cdot 10^{-10}N\cdot nm,$$
$$D'' = 2h(\gamma-\varepsilon) = \gamma^* - \varepsilon^* = -0,91\cdot 10^{-10}N\cdot nm,\ \nu_m = 2\frac{D''}{D'} = -0,41,\ \nu = 0. \tag{14.25}$$

14.4 The Problem of Static Transverse Bending of a Rectangular Graphene Sheet

Let's consider an example of a static transverse bending of a graphene sheet, when the sheet occupies the region of a rectangle: $0 \le x \le a_1,\ 0 \le y \le a_2$. It is assumed that the contour of graphene is hinged supported. For the boundary conditions we have:

$$x = 0, a_1 : w = 0,\ L_{12} = 0,\ \Omega_1 = 0;$$
$$y = 0, a_2 : w = 0,\ L_{21} = 0,\ \Omega_2 = 0. \tag{14.26}$$

Let's consider the loading of a graphene sheet with a transverse load of the form:

$$p_3(x,y) = p_0 \sin\frac{\pi x}{a_1}\sin\frac{\pi y}{a_2}. \tag{14.27}$$

The solution of the system of Eqs. (14.13) is set in the form:

$$w(x,y) = w_0 \sin\frac{\pi x}{a_1}\cdot\sin\frac{\pi y}{a_2},$$
$$\Omega_1 = \Omega_{10}\sin\frac{\pi x}{a_1}\cdot\cos\frac{\pi y}{a_2}, \tag{14.28}$$
$$\Omega_2 = \Omega_{20}\cos\frac{\pi x}{a_1}\cdot\sin\frac{\pi y}{a_2}.$$

Here the multipliers $w_0, \Omega_{10}, \Omega_{20}$ are to be determined. It should be noted that with the help of (14.28) the set boundary conditions (14.26) will be satisfied.

To determine $w_0, \Omega_{10}, \Omega_{20}$, (14.28) should be substituted into the system of Eqs. (14.13), as a result, we come to the solution of an algebraic linear inhomogeneous system of equations. If $p_0 = 10^6\frac{N}{m^2}$, $a_1 = a_2 = 20nm$, then for the maximum deflection of the graphene sheet we'll obtain:

$$w_0 = \frac{p_0 a_1^2}{2\pi^2\tilde{D}'}\left(1 + \frac{\tilde{D}'}{D_*}\right) = 1,17nm,$$

where $\tilde{D}' = \frac{D'}{a_1^2} \cdot 2\pi^2$.

14.5 The Problem of Stability of the Initially Compressed State of a Graphene Sheet. Determination of the Critical Load

Let us consider a rectangular plate (graphene sheet), in which all sides are supported; normal compressive forces are uniformly distributed on the two sides $x = 0, x = a_1$ (the magnitude of the force per unit length). It is required to determine the critical value of forces p_{cr}, at which the flat form of equilibrium ceases to be the only and stable form (the plate loses its stability).

Before the loss of stability, the stressed state of the plate is flat. For this state, it is necessary to consider the static equilibrium Eqs. (14.1) $\left(p_1^{\pm} = p_2^{\pm} = m_3^{\pm} = 0\right)$ and the boundary conditions: $x = 0, a_1, T_{11} = -p, S_{21} = 0, L_{23} = 0; y = 0, a_2, T_{22} = 0, S_{12} = 0, L_{13} = 0$.

This boundary value problem in the region of the middle plane of the plate: $0 \le x \le a_1, 0 \le y \le a_2$, has a solution: $T_{11}(x, y) = p, T_{22}(x, y) \equiv 0, S_{12}(x, y) \equiv 0, S_{21}(x, y) \equiv 0, L_{13}(x, y) \equiv 0, L_{23}(x, y) \equiv 0$.

The system of equations for a curved surface will be:

Equilibrium equations:

$$\begin{gathered}
\frac{\partial N_{13}}{\partial x} + \frac{\partial N_{23}}{\partial y} = p\frac{\partial^2 w}{\partial x^2}, \\
\frac{\partial L_{11}}{\partial x} + \frac{\partial L_{21}}{\partial y} + N_{23} = 0, \\
\frac{\partial L_{12}}{\partial x} + \frac{\partial L_{22}}{\partial y} - N_{13} = 0.
\end{gathered} \tag{14.29}$$

Elasticity relations will be expressed by Eqs. (14.9), geometric relations- by (14.10).

The system of Eqs. (14.29), (14.9), (14.10), can be introduced as follows:

$$\begin{cases}
\Delta w + \dfrac{\partial \Omega_2}{\partial x} - \dfrac{\partial \Omega_1}{\partial y} = \dfrac{p}{D_*}\dfrac{\partial^2 w}{\partial x^2}, \\
\Delta \Omega_1 + \nu_m\left(\dfrac{\partial^2 \Omega_1}{\partial x^2} + \dfrac{\partial^2 \Omega_2}{\partial x \partial y}\right) + \dfrac{D_*}{D'}\left(\dfrac{\partial w}{\partial y} - \Omega_1\right) = 0, \\
\Delta \Omega_2 + \nu_m\left(\dfrac{\partial^2 \Omega_2}{\partial y^2} + \dfrac{\partial^2 \Omega_1}{\partial x \partial y}\right) - \dfrac{D_*}{D'}\left(\dfrac{\partial w}{\partial x} + \Omega_2\right) = 0.
\end{cases} \tag{14.30}$$

It is necessary to find a nonzero solution of the system of homogeneous Eqs. (14.30) that satisfies the boundary conditions on the sides of the plate.

In the case of hinged support boundary conditions (14.26), as in the previous case, after introducing the functions $\phi(x, y)$ and $\psi(x, y)$ according to formulas (14.19), the system of Eqs. (14.30) is reduced to solve the following equations by w:

$$\Delta\Delta w + \frac{p}{D'} \cdot \frac{\partial^2 w}{\partial x^2} - \frac{p}{D_*} \cdot \frac{\partial^2(\Delta w)}{\partial x^2} = 0. \tag{14.31}$$

The solution of this equation will be introduced as follows:

$$w = w_0 \sin\frac{m\pi x}{a_1} \cdot \sin\frac{n\pi y}{a_2}, \tag{14.32}$$

where w_0 is constant coefficient, m and n are integer values.

Substituting expression (14.32) into Eq. (14.31), considering that we are interested in a solution that is not equal to zero, we obtain all the values p corresponding to the values $m = 1, 2, 3, ..., n = 1, 2, 3, ...,$ at which the curvature of the form (14.32) becomes possible. From the entire set of values p, it is necessary to choose the smallest, it will be critical.

If, in particular, we consider a square plate $a_1 = a_2$ when $m = n = 1$, for the critical load we obtain the formula

$$p_{cr} = \frac{\frac{4\pi^2 D'}{a_1^2}}{1 + \frac{D'}{D_*} \cdot \frac{2\pi^2}{a_1^2}}.$$

If $a_1 = 20nm$, then $p_{cr} = 0,04\frac{N}{m}.p_{cr} = 0,04\frac{N}{m}$.

14.6 The Problem of Natural Transverse-Flexural Vibrations of a Graphene Sheet

We'll consider the system of dynamic equations for transverse bending of the moment-membrane theory of elastic plates (8)-(10). The resolving system of equations for natural transverse bending vibrations of elastic plates according to the moment-membrane theory is represented as follows

$$\begin{cases} \Delta w + \left(\dfrac{\partial \Omega_2}{\partial x} - \dfrac{\partial \Omega_1}{\partial y}\right) = \dfrac{\rho_0}{D_*}\dfrac{\partial^2 w}{\partial t^2}, \\ \Delta \Omega_1 + \eta \dfrac{\partial}{\partial x}\left(\dfrac{\partial \Omega_1}{\partial x} + \dfrac{\partial \Omega_2}{\partial y}\right) + \dfrac{D_*}{D'}\left(\dfrac{\partial w}{\partial y} - \Omega_1\right) = \dfrac{I_0}{D'} \cdot \dfrac{\partial^2 \Omega_1}{\partial t^2}, \\ \Delta \Omega_2 + \eta \dfrac{\partial}{\partial y}\left(\dfrac{\partial \Omega_1}{\partial y} + \dfrac{\partial \Omega_2}{\partial y}\right) - \dfrac{D_*}{D'}\left(\dfrac{\partial w}{\partial x} + \Omega_2\right) = \dfrac{I_0}{D'} \cdot \dfrac{\partial^2 \Omega_2}{\partial t^2}. \end{cases} \quad (14.33)$$

This system of equations can be reduced to the following equivalent form:

$$\begin{aligned} &\Delta\Delta w - \left(\frac{\rho_0}{D_*} + \frac{I_0}{D'}\right)\frac{\partial^2(\Delta w)}{\partial t^2} + \frac{\rho_0}{D'}\frac{\partial^2 w}{\partial t^2} + \frac{I_0 \cdot \rho_0}{D_* \cdot D'} \cdot \frac{\partial^4 w}{\partial t^4} = 0, \\ &\Delta\psi - \frac{D_*}{D'(1+2\eta)}\psi = \frac{I_0}{D'(1+2\eta)} \cdot \frac{\partial^2 \psi}{\partial t^2}. \end{aligned} \quad (14.34)$$

Let all sides of the rectangular plate be hinged supported (in this case $\psi(x, y, t) \equiv 0$), then we have to consider only the first equation from system (14.34), the solution of which can be represented in the form (boundary conditions are satisfied):

$$w(x, y, t) = w_{mn} \sin \frac{m\pi x}{a} \sin \frac{n\pi y}{b} e^{i\omega t}. \quad (14.35)$$

The frequency order of the lowest form of natural vibrations (in the case of a square plate) can be determined by the following approximate formula

$$\omega \approx \frac{2\pi^2}{a^2}\sqrt{\frac{D'}{\rho_0}}. \quad (14.36)$$

Here $\rho_0 = 0,77 \cdot 10^{-6} \frac{kg}{m^2}$, $D' = 4,15 \cdot 10^{-10} N \cdot nm$.

If $a = 10$, then for ω we'll obtain $\omega \approx 0,02 \cdot 10^{12} Hz$, and if $a = 5nm$, we'll have: $\omega \approx 0,1 \cdot 10^{12} Hz$.

14.7 Conclusion

The models of plane stress state and transverse bending of the moment-membrane theory of plates as continuous models of graphene sheet deformations are introduced in the paper. The following three applied problems are studied for a rectangular graphene sheet: (1) the problem of its static transverse bending (the maximum displacement is determined), (2) the problem of the stability of the initially compressed state of the graphene sheet (the critical value of the compressive force is determined), (3) the problem of natural transverse-bending vibrations of the graphene sheet, determining the spectrum of its natural vibrations and showing that the lowest vibration frequency is in the giga- terahertz (GHz, THz) region.

Acknowledgments The Work Was Supported by the Science Committee of RA, in the Frames of the Research Project №. 21 T-2C093.

References

1. Panin VE, Grinyaev YuV, Egorushki VE (2010) Foundations of physical mesomechanics of structurally inhomogeneous media. Mech Solids 45:501–518
2. Ivanova EA, Krivtsov AM, Morozov NF, Firsova AD (2003) Inclusion of the moment interaction in the calculation of the flexural rigidity of nanostructures. Dokl Phys 48:455–458
3. Ivanova EA, Krivtsov AM, Morozov NF (2007) Derivation of macroscopic relations of the elasticity of complex crystal lattices taking into account the moment interactions at the microlevel. Appl Math Mech 71:543–561
4. Berinsky IE, Krivtsova AM : and etc; under total ed, (2014) Modern problems of mechanics. Mechanical properties of covalent crystals Publishing house of Politekhn. university, Loboda, O.S, St. Petersburg
5. Sargsyan SH (2022) Beam and continual models for deformations of two-dimensional nanomaterials. Phys Mesomech 25:109–121
6. Sargsyan SH (2020) A thin shell model within the moment theory of elasticity with the concept of deformation by shear plus rotation. Phys Mesomech 23:309–315
7. Sargsyan SH (2022) Variation principles of moment-membrane theory of shells. Moscow Univ Mech Bull 77:1–11
8. Morozov NF (1984) Mathematical problems of the theory of cracks. M.: Nauka
9. Pelekh BL (1977) Stress concentration near hole during bending of transversely isotropic plates. Kyiv: Nauka dumka

Chapter 15
Fracture Mechanisms of Hydrogen-Charged Metal Samples During a Three-Point Bending Test

Yuliya S. Sedova

Abstract The effect of hydrogen saturation of steel specimens on the results of their standardized testing for hydrogen cracking resistance was studied. When modeling the three-point bending of the sample, the hydrogen skin effect is taken into account, which is observed during standardized saturation of the samples with hydrogen in various electrolyte solutions. The classical decohesion HEDE model of hydrogen embrittlement is used. It is shown that, despite the microscopic depth, the skin effect leads to a dual nature of destruction, when both areas of hydrogen embrittlement and areas of ordinary fracture are observed at the fracture of the sample. Calculation results showed a strong influence of the hydrogen skin layer on the results of standardized testing of metals. This skin effect has a significant impact in the propagation of the destruction process along a metal sample and should be taken into account when performing industrial tests, simulations and experimental studies.

15.1 Introduction

During the production of metals and parts from them, the accumulation of the so-called metallurgical dissolved hydrogen occurs. It adversely affects all mechanical and technological characteristics of metals, is one of the main causes of brittleness, porosity and the formation of flocs in rolled products. The concentration of metallurgical hydrogen at the plants is controlled. In steels and aluminum alloys it is maintained at a level of the order of 1 ppm or less.

The second source of dissolved hydrogen is the aggressive environment in which metal parts are operated. Any corrosion process is accompanied by the release of hydrogen and its absorption inside the metal. External mechanical loads and the presence of pure hydrogen in the environment lead to a significant acceleration of the hydrogen degradation of metal properties.

Y. S. Sedova (✉)
Institute for Problems in Mechanical Engineering Russian Academy of Sciences (IPME RAS)
Saint Petersburg, Saint Petersburg, Russia
e-mail: sedova.yus@mail.ru

H. Altenbach and V. Kuzkin (eds.), *Advanced Problems in Mechanics*,
https://doi.org/10.1007/978-3-031-80103-7_15

Over the past few decades, a system of standardized testing for hydrogen-induced cracking (HIC) has been developed and implemented in industry [1–3]. As a rule, the procedure consists of exposure (hydrogen saturation) of metal samples in a corrosive environment, which lasts up to 100 h, mechanical tests, as well as examination of the fracture surfaces of the samples under a microscope.

The use of these tests has made it possible to achieve high resistance to HIC of modern pipe steels. If twenty years ago, hydrogen-induced cracks formed inside steel specimens during testing [4], then modern steels are characterized by only insignificant external microcracking. At the same time, the mechanical characteristics of the samples noticeably deteriorate after saturation with hydrogen, and high-strength steels are characterized by a multiple decrease in crack resistance [5, 6]. Numerous fractographic studies [7–10], carried out during an experimental study of steels fractured during mechanical tests, indicate the presence of two characteristic fracture regions in the samples surface: brittle and ductile. Modern alloys are widely used in various fields of energy and industry. Therefore, the effect of their rapid degradation upon saturation with hydrogen must be investigated and minimized.

The average partial concentration of hydrogen atoms to the atoms of the matrix in the metal in the process of HIC-testing increases to values no more than 1:30,000 [11]. In addition, the results of recent studies show that in aqueous corrosive solutions and at cathodic polarization, a skin effect of hydrogen saturation occurs. It lies in the fact that all the hydrogen absorbed by the metal is concentrated in a thin surface layer of samples with a thickness of about 50–100 μm [12–14]. With a diameter of the working part of the sample of 8–10 mm, this layer can be classified as layers of small thickness (the ratio of 1:100 is typical for the theory of thin shells). Thus, from the point of view of mechanics, we are faced with the problem of the influence of a small parameter.

To date, several approaches have been proposed to describe hydrogen-induced fracture.

One of the first was the hydrogen-enhanced decohesion (HEDE) model [15]. It describes brittle fracture as a result of the accumulation of hydrogen at the crack tip and the development of hydrogen embrittlement without plastic deformation. The idea of an approach at the microscopic level is based on the fact that interstitial hydrogen expands the metal lattice of crystals, thereby contributing to a decrease in the adhesion strength of atoms [16]. From an energetic point of view, hydrogen atoms lower the energy barrier to destruction, which leads to grain separation, or decohesion.

A fundamentally different approach to modeling the hydrogen degradation of the mechanical properties of materials is considered in the model of hydrogen-enhanced localized plasticity (HELP) [17, 18]. Within the framework of this model, hydrogen dissolved in the metal is interpreted as a chemical substance that reduces the energy required for the generation of dislocations. It is assumed that at the crack tip, where hydrogen is concentrated under the action of internal stresses, the metal "softens", or localized plasticity occurs.

The dual nature of fracture observed during HIC-testing is usually interpreted using a combination of both models: HELP + HEDE. It is assumed that if the local concentrations of hydrogen at the crack tip are above a certain "switching threshold", then the brittle HEDE mechanism operates, and if the concentration is below the threshold, the HELP mechanism acts. One of the problems of this approach is the exponential (up to 100 times) growth of the hydrogen concentration during plastic deformation, which directly follows from the equations of the HELP model [19]. Thus, with successive "adding" of models, localized plastic deformation initiated by hydrogen should lead to a multiple increase in its local concentration and the triggering of a brittle decohesion fracture mechanism.

In addition, the HELP model actually describes an increase in ductility due to a decrease in the material's yield strength. However, the bulk of the experimental data does not reveal a decrease in the yield strength during testing of samples [6]. All this leads to the fact that the idea that the HELP model begins to act and determines the main fracture cross section does not correspond to the existing experimental and research data.

These HELP + HEDE paradoxes allow us to opt for the decohesion model of hydrogen destruction.

15.2 The HEDE Model of Brittle Fracture

The modern concept of the HEDE model consists of the following main elements: (1) equations of an elastic or viscoelastic continuum (including for materials with hardening), (2) diffusion equations describing the redistribution and accumulation of hydrogen, and (3) dependencies describing the hydrogen degradation of the parameters of the cohesion law.

Hydrogen transport inside a solid was traditionally considered as a diffusion process, and the Fick's law [20] was used to describe it:

$$\frac{\partial C}{\partial t} = \nabla.[D(T,\boldsymbol{r})\nabla C], \tag{15.1}$$

where C is the hydrogen concentration, D—the diffusion coefficient, T—the absolute temperature, $\boldsymbol{r}$—the radius vector, t—the time and $\nabla = \boldsymbol{i}_k \frac{\partial}{\partial x_k}$—the Del operator.

However, modern approaches, as a rule, are based on the introduction of terms that consider the thermodynamic or chemical potential $\mu = \mu_0 + RT\ln C - V_H p$ [21]. In this case, Eq. (15.1) takes the form

$$\frac{\partial C}{\partial t} = \nabla.\left[D(T,\boldsymbol{r})\left(\nabla C - \frac{CV_H}{RT}\nabla p\right)\right], \tag{15.2}$$

and taking into account the influence of mechanical stresses, it has the following form:

$$\frac{\partial C}{\partial t} = D\nabla^2 C - \frac{DV_H}{RT}\nabla C.\nabla p - \frac{DCV_H}{RT}\nabla^2 p, \tag{15.3}$$

where p is the hydrostatic pressure determined by the spherical part of the stress tensor, V_H—the partial molar volume of hydrogen [22].

According to Gorsky's law [23], hydrogen tends to regions of maximum tensile stresses. An increase in its concentration in these areas inevitably leads to a decrease in the adhesion of the crack edges, resulting in decohesion [15]. To describe this phenomenon within the framework of the model the parameter of the degree of filling of the free surface of the crack edges with hydrogen atoms θ is used:

$$\theta = \frac{C}{C + \exp\left(-\frac{\Delta g_H}{RT}\right)}, \tag{15.4}$$

where Δg_H is the difference between the Gibbs free energy for hydrogen between the adsorbed inside the crystal lattice and the free state (established by experiment).

It is important that this expression assumes the value of the partial concentration of hydrogen. Therefore, in the case of presenting the content of a substance in ppm, the following ratio should be used.

$$\theta = \frac{55.85.C.10^{-16}}{55.85.C.10^{-16} + \exp\left(-\frac{\Delta g_H}{RT}\right)} \tag{15.5}$$

This expression was proposed by Serebrinsky [24] based on the thermodynamic ratio of the partial concentration of hydrogen inside and on the surface of a monocrystal, obtained by McLean [25].

Using the value of the parameter θ, one can calculate the specific energy of the free surface $\gamma(\theta)$, which depends on the sorption of hydrogen on it. To describe this relationship, most existing studies use the formula proposed by Serebrinsky [24]:

$$\gamma(\theta) = \left(1 - 1.0467\theta + 0.1687\theta^2\right)\gamma(0), \tag{15.6}$$

where $\gamma(0)$ is the surface energy in the absence of hydrogen.

Further, within the framework of the brittle fracture model, the energetical identity is considered

$$G = 2\gamma(\theta) = f\left(\sigma_{cr}^{coh}, \delta_{cr}\right), \tag{15.7}$$

in which G is the fracture energy, $f\left(\sigma_{cr}^{coh}, \delta_{cr}\right)$ is a linear function depending on the value of the maximum cohesive stresses σ_{cr}^{coh}, normal to the crack edges, and on the maximum value of the relative displacement of the crack edges δ_{cr}, at which the bonds between them do not break. The form of the function $f\left(\sigma_{cr}^{coh}, \delta_{cr}\right)$ is determined by the establishment of the cohesive law.

Based on the assumption that the value of δ_{cr} weakly depends on the value of θ, the hydrogen degradation law is postulated:

$$\sigma_{cr}^{coh}(\theta) = \left(1 - 1.0467\theta + 0.1687\theta^2\right)\sigma^{coh}(0) \tag{15.8}$$

where $\sigma^{coh}(0)$ are the cohesive stresses normal to the crack edges, acting in the absence of hydrogen.

According to the criterion of crack initiation and propagation in terms of the HEDE decohesion mechanism, a crack retains its configuration until the level of elastic stresses normal to its edges and acting near the crack tip does not exceed the cohesive stresses that bind them together. As soon as the hydrogen accumulated at the crack tip under the action of the chemical potential reduces these stresses to such an extent that this condition is violated, the crack growth will be initiated, accompanied by the formation of a new free surface. This shift, at the same time, will cause a local increase in the level of normal stresses in the new position of the crack tip, which will entail an increase in the local concentration of hydrogen. There is another decrease in the value of cohesive stresses until the fracture stability condition is violated, and the process is repeated again and again.

On the other hand, the energy required for destruction can be expressed in terms of material parameters. For a normal separation crack, it can be written as

$$G = \frac{K_{Ic}^2}{\hat{E}}, \tag{15.9}$$

where K_{Ic} is the critical stress intensity factor (SIF) acting near the crack tip; $\widehat{E} = E/\left(1 - \nu^2\right)$ or $\widehat{E} = E$ for the plane strain case or for the plane stress case, respectively (E is Young's modulus, v is Poisson's ratio).

Considering identity (7), we can show that a decrease in the specific energy of the free surface caused by an increase in the hydrogen concentration also leads to a decrease in K_{Ic}:

$$\frac{K_{Ic}^2}{K_{Ic}^2(0)} = \frac{\gamma(\theta)}{\gamma(0)} = 1 - 1.0467\theta + 0.1687\theta^2.$$

Thus, the value of the critical stress intensity factor (the fracture toughness of the material) depends on the hydrogen content as

$$K_{Ic}(\theta) = \sqrt{1 - 1.0467\theta + 0.1687\theta^2}K_{Ic}(0), \tag{15.10}$$

where $K_{Ic}(0)$ is the experimentally established value of the fracture toughness in the absence of hydrogen.

Hence, it follows that the presence of hydrogen dissolved in the material introduces a correction into the Irwin's force fracture criterion ($K_I = K_{Ic}$) [26, 27], so crack propagation becomes possible at a lower value of K_I.

15.3 Finite Volume Simulation

Within the framework of this study, the problem of three-point bending of a single edge notched bend (SENB) specimen saturated with hydrogen using the HEDE brittle fracture model was considered. We excluded the solving with the standard finite element packages with an inbuilt cohesive zone modeling subroutine and a crack propagation module. This is since using them involves introducing special fictitious cohesive elements into consideration and working with a certain cohesive law. Such an introduction requires setting many additional model parameters and accepting several additional assumptions.

The model was implemented using our own developed software algorithm in the Microsoft Visual Studio environment, written in C++ and based on the finite volume method.

15.3.1 Numerical Procedure

The computational procedure consisted of two successive stages: (i) the solution of the problem of the stress–strain state of the loaded body; (ii) the solution of the diffusion problem of the redistribution of hydrogen over the sample, within which the cohesive stresses σ_{Zc} and the fracture toughness K_{Ic} values were also calculated.

At the stage (i) the system of equations of the linear theory of elasticity (11–13) was solved. It includes the differential equation of equilibrium (11), the definition of the linear strain tensor (12) and the generalized Hooke's law (13).

$$\begin{cases} \nabla \cdot \sigma + f = 0, \\ \varepsilon = \frac{1}{2}(\nabla \boldsymbol{u} + \boldsymbol{u}\nabla), \\ \sigma = \lambda \theta E + 2G\varepsilon, \end{cases} \tag{15.11}$$

where $\boldsymbol{\sigma}$—the Cauchy stress tensor; $\boldsymbol{f}$—the intensity of body forces; $\boldsymbol{u}$—the displacement vector; λ and G—the Lame parameters, $\theta = \mathrm{tr}(\boldsymbol{\varepsilon})$—the relative volume change, the hydrostatic pressure $p = \frac{1}{3}\mathrm{tr}(\boldsymbol{\sigma})$.

For the numerical solution of Eq. (15.11), the method of establishment proposed by Wilkins [28] was used.

The characteristics of the stress–strain state of the sample, obtained at the (i) step of the procedure, were transferred to the stage (ii). It included the solution of the system of Eqs. (15.3), (15.5), (15.8) and (15.10) described earlier:

$$\begin{cases} \frac{\partial C}{\partial t} = D\nabla^2 C - \frac{DV_H}{RT}\nabla C \cdot \nabla p - \frac{DCV_H}{RT}\nabla^2 p, \\ \theta = \frac{55.85.C.10^{-6}}{55.85.C.10^{-6} + \exp\left(-\frac{\Delta g_H}{RT}\right)}, \\ \sigma_{cr}^{coh}(\theta) = \left(1 - 1.0467\theta + 0.1687\theta^2\right)\sigma^{coh}(0), \\ K_{Ic}(\theta) = \sqrt{1 - 1.0467\theta + 0.1687\theta^2} K_{Ic}(0). \end{cases}$$

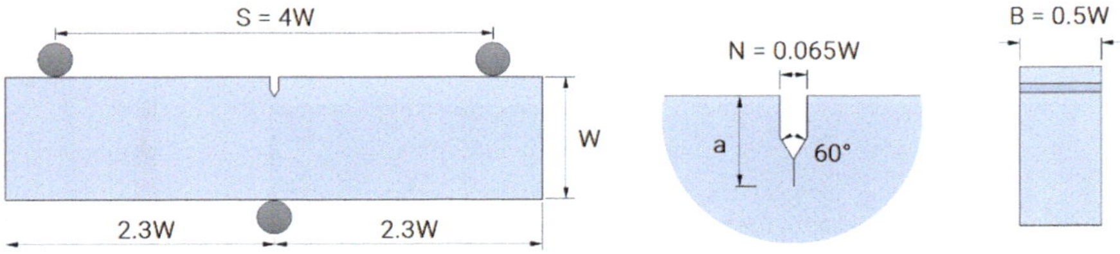

Fig. 15.1 Single edge notched bend (SENB) specimen

The numerical solution of Eq. (15.3) was also performed based on the Wilkins establishment method [28].

To calculate the current value of the SIF in the vicinity of the crack tip, Srawley's approximate formula [29] was used:

$$\begin{aligned} K_I &= \frac{3}{2}\frac{S\sqrt{W}P\sqrt{\alpha}}{WB}F, \\ F &= \frac{1.99-\alpha(1-\alpha)\left(2.15-3.93\alpha+2.7\alpha^2\right)}{2(1+2\alpha)(1-\alpha)^{\frac{3}{2}}}, \alpha = \frac{a}{W}, \end{aligned} \tag{15.14}$$

where W and B—the sample width and thickness, respectively, P—the applied load, S—the loading span, a—nominal crack length, (Fig. 15.1).

The results of the stage (ii) determined the current state and possibility of crack growth in accordance with the criterion of cohesive stresses or Irwin's force fracture criterion.

In order to be able to detect a possible ductile fracture and describe the elastic–plastic behavior of the material, we included a module in our program code that takes into account the fluidity of the body. This subroutine is based on the approach outlined in the paper [30] and is a modification of the method, proposed by Wilkins [31].

15.3.2 Initial Conditions and Parameters of the Model

In this study we performed a numerical simulation of a three-point bending of a sample with a thickness of $W = 20mm$ (Fig. 15.1).

Due to the symmetry, the geometry of the simulated area was half of the sample under study. On the computational domain we formed a non-orthogonal mesh consisting of quadrangular structural elements, which, in turn, were divided into two triangular ones. The finite volume model is shown in Fig. 15.2 (the symmetry line runs lengthwise the left side of the area along the direction of the Oy axis). The prohibition of displacements along the horizontal direction Ox was set on the nodes of the left edge. The calculation was carried out in a two-dimensional plane formulation.

Fig. 15.2 Finite volume model, initial and boundary conditions

The material of the sample under consideration is X70 pipe steel with the following physical properties: ultimate strength $\sigma_B = 620MPa$, yield stress $\sigma_Y = 420MPa$, mass density $\rho = 7800kg/m^3$, shear modulus $G = 80.7GPa$, the critical stress intensity factor $K_{Ic}(0) = 138.5MPa\sqrt{m}$. In addition, the diffusion coefficient $D = 2.5 \cdot 10^{-11} m^2/s$, the partial molar volume of hydrogen $V_H = 2 \cdot 10^{-6} m^3/mol$[32], the difference in the Gibbs free energy for hydrogen between the hydrogen adsorbed inside the crystal lattice and the free state $g_H = 30kJ/mol$ [32], and the absolute temperature $T = 298°K$. The value $\sigma^{coh}(0)$ was set equal to $1.5\sigma_Y$.

We solved the problem for two cases: for a hydrogen-free body and for a sample saturated with hydrogen. For the last of these, we have specified the following initial distribution of hydrogen concentration over the sample. In the surface layer with a thickness of one structural element, the content of the substance was set equal to $10ppm$, representing the experimentally observed skin effect of hydrogen saturation. For the rest of the model volume, we took into account a uniform background concentration of $0.2ppm$.

We simulated crack initiation and propagation as follows. After obtaining the characteristics of the stress–strain state of the body, at the stage (ii) of the computational procedure, the analysis of the diffusion problem on the redistribution of hydrogen over the sample was started. During the analysis, the status of two inequalities was monitored: for stresses $\sigma^{el} \leq \sigma^{coh}_{cr}$ and for stress intensity factors $K_I \leq K_{Ic}(\theta)$ at the nodes of the mesh along the assumed crack propagation line (along the left side of the area). As soon as one of these inequalities was violated in a node, caused by an increase in the hydrogen concentration or an increase in the level of elastic stresses, the restriction on movement along the horizontal axis Ox was removed in this node, thereby simulating a break in the connection with the node on the opposite side of the crack (that is decohesion). And then the node got the opportunity to move freely under the current load.

Thus, the approach we have chosen using the diffusion Eq. (15.3), the definition of the parameter (5), and the hydrogen degradation laws (8), (10) makes it possible to build a recurrent scheme for solving the problem of hydrogen embrittlement. First, it is necessary to set the initial distribution of the hydrogen concentration over the area under study, corresponding to the result of saturation of the samples with hydrogen, and obtain a solution to the problem of the stress–strain state of the body on the stage (i). Next, calculate the diffusion problem of hydrogen redistribution—stage (ii), taking into account mechanical stresses, while simultaneously monitoring the level of cohesive stresses and fracture toughness values that determine the crack configuration. Then return to the stage (i) of the task and repeat the procedure. According to the HEDE model, the distribution of hydrogen does not directly affect the stress state of the sample. Due to the fact that the process of diffusion and redistribution of matter requires some time, the recurrent scheme can be considered as temporary, and by setting a certain value of the iteration step, we can estimate the incubation time for the initiation and propagation of a crack in the body.

15.4 Simulation Results

It was found that when a sample unsaturated with hydrogen is loaded, the initiation of a crack at the notch tip is accompanied by the formation of a local zone of plastic deformation in it (Fig. 15.3a).

As the destruction spreads, irreversible changes take on a large value and cover more and more area of the material. The evolution of the plastic deformation zone is shown in Fig. 15.3a–c.

Thus, all in the same key the autocatalytic ductile fracture propagated up to the complete rupture of the sample. Thereby the data of the performed verification calculation showed the expected result corresponding to a standard loading condition.

Let us now turn to the results of numerical simulation of the three-point bending of a sample saturated with hydrogen.

It was found that the destruction of a body containing hydrogen begins at a noticeably lower load level (at a 0.75 MPa). It starts with the immediate crack initiation at the notch tip by a brittle decohesion mechanism at the very first step in time integration. This is explained by the high content of hydrogen in the surface layer.

The incubation time for the initiation and propagation of a crack is shown on the Fig. 15.4. It, in fact, is the time between the fulfillment of the fracture criterion in two successive nodes of the finite volume model.

There are several main stages in this graph. After instantaneous destruction initiation at the first two mesh nodes, further the crack develops according to a brittle decohesion mechanism with an inherent propagation incubation period, during which hydrogen is redistributed and comes to the current crack tip.

After that, the level of elastic stresses acting near the concentrator becomes decisive, hereupon further crack propagation becomes possible without the participation

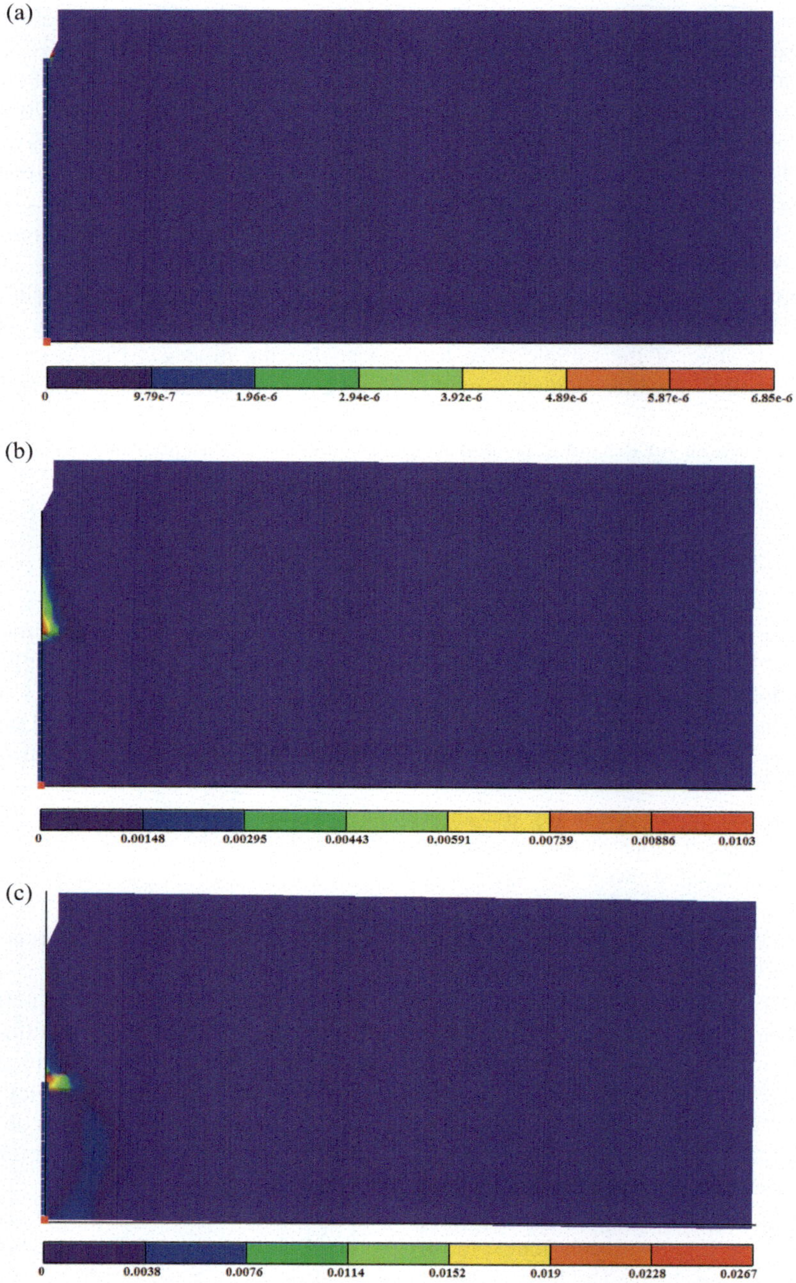

Fig. 15.3 Evolution of the plastic deformation zone during fracture propagation: **a** emerging of plastic deformation at a 1.5 MPa load, the crack is not initiated; **b** development of the plastic deformation zone (at a 2.5 MPa load); **c** development of an extensive plastic deformation zone (at a 2.5 MPa load)

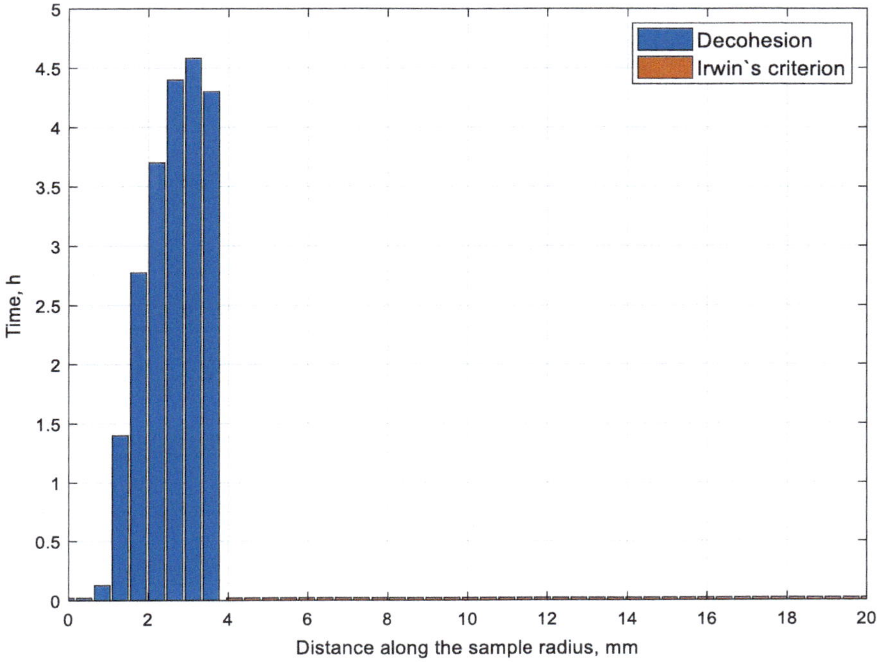

Fig. 15.4 The incubation time for the initiation and propagation of a crack

of hydrogen. Moreover, when the destruction propagates 4*mm* deep into the sample, the acting stresses become so high that the fracture mechanism changes from brittle decohesion to that triggered by exceeding the critical SIF. Besides we found out that the size of the region of hydrogen embrittlement is determined only by the level of the applied external loading (the higher it is, the smaller the zone size).

And only when the crack reaches a certain length, a plastic deformation zone is formed. At an early stage, it is very localized and concentrated at the crack tip. The subsequent destruction of the sample is inseparably accompanied by the growth of this region and an increase in the irreversible deformation of the body (Fig. 15.5).

Further depending on the properties of the material and the loading conditions, it can develop as destruction advances and lead to a change from the brittle fracture mechanism to the ductile one.

Thus, we have established that while an unsaturated body during a three-point bending test demonstrates an unambiguous fracture accompanied by the development of plastic deformations, the fracture surface of a hydrogenated sample has a complex, composite character. The non-occurrence of plastic deformations at the first stage of crack development in such a body is associated with an explicit hydrogen effect.

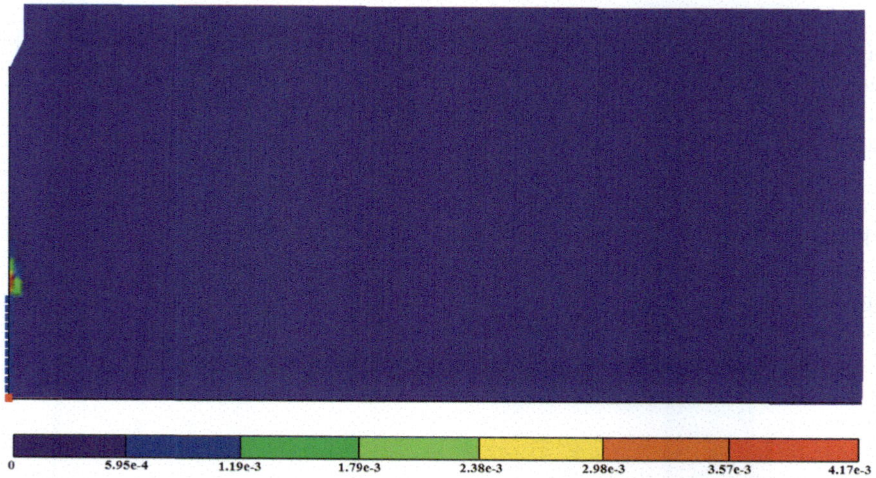

Fig. 15.5 Formation of a plastic deformation zone in a sample saturated with hydrogen

15.5 Discussion

We have obtained the results of a finite volume solution of the equations of the classical model of hydrogen-induced brittle fracture HEDE for a standard three-point bending test considering the skin-effect of saturation of sample with hydrogen.

For calculations, we used experimentally measured values of the content of the substance, obtained during saturation of metal parts with hydrogen. And the solution of this problem led us, at first sight, to a paradoxical result.

The destruction of a sample that does not contain hydrogen, as expected, was possible at a noticeable level of external load, when a plastic deformation zone is already formed at the notch tip. Accordingly, the initiation and growth of a crack in it is inseparably accompanied by the development of plastic deformation, which covers an increasingly large area around its edges, and is supposed to occur according to the ductile mechanism.

At the same time, the initiation and growth of a primary crack in a sample saturated with hydrogen with a characteristic skin layer begins at a significantly lower level of bending load, when mechanical testing is carried out in the region of elastic deformations. The initial destruction occurs here by a brittle decohesion mechanism associated with an increased concentration of the substance. However, further crack propagation will be accompanied by an increase in the level of stresses acting in the material, because of which a local region of plastic deformations may appear near its tip, which can form even at a distance of about 1/3 of the sample thickness, as was shown in our study. And further, with a sufficient degree of development of the plastic deformation zone in the vicinity of the concentrator, we also have the right to expect a transition to ductile fracture of the material. Such duality is often encountered in experimental works [8, 9, 33, 34].

In addition, we carried out calculations with varying skin layer thickness, hydrogen concentration and applied load. They showed that the width of the hydrogen embrittlement area significantly depends only on the level of the external load applied to the sample (or the sample tensile rate); the larger it is, the smaller the area size, what also qualitatively corresponds to the experimental data [8].

It is also necessary to pay attention to the following fact established by us. Calculation results show that when the mechanical load is applied the microcrack in the skin layer itself is formed immediately according to the brittle decohesion mechanism. This phenomenon is also consistent with the numerous experimental data on the occurrence of microscopic cracks (so-called "quasi-cleavages") on the surface of charged metal samples upon deformation.

Thus, the classical models used by us make it possible to describe the critical influence of a skin-effect small in relation to area volume (the influence of a small parameter) on the destruction of samples and to correlate the experimentally observed dual nature of the metals fracture with the procedure for their artificial saturation with hydrogen. The obtained results explain the differences between the results of testing steel parts and their operation in aggressive environments.

15.6 Conclusions

We performed a finite volume solution of the of the destruction of a single edge notched bend (SENB) specimen saturated with hydrogen. The classical mechanism of hydrogen-enhanced decohesion HEDE is used as a mechanical model of hydrogen embrittlement. The experimentally established values of the content of the substance, obtained during saturation of metal parts with hydrogen, was considered.

The calculation results showed that in this case, destruction begins at significantly lower loads from the surface of the sample, as a brittle fracture induced by hydrogen. Then, when the crack reaches a certain length and due to the "lag" of diffusion flows, it will grow by itself at ordinary background values of the hydrogen concentration.

Such a situation will correspond to a dual nature on the fracture pattern in experimental practice. The specimen failure will be represented as areas of hydrogen embrittlement interspersed with areas of common destruction corresponding to the classical theory of crack growth.

In this way the experimentally established non-uniform distribution of hydrogen concentration after saturation of the samples can be the genuine cause of the dual nature of fracture, which is currently supposed to be explained by a hybrid model of the simultaneous action of HELP and HEDE mechanisms at the crack tip.

The artificial saturation of samples with hydrogen, as a rule, differs significantly in the distribution of concentrations from the natural saturation with this substance occurring during operation. This fact must be considered, both when testing metals for resistance to hydrogen cracking, and in experimental studies, since the skin effect occurs when using standard hydrogen saturation methods and exactly strongly affects the results of the tests provided for by international standards. Moreover, the obtained results indicate the need for reverification and possible changes in the methodology for implementation such tests.

References

1. ISO11114-4 (2017). Transportable gas cylinders—compatibility of cylinder and valve materials with gas contents—part 4: test methods for selecting metallic materials resistant to hydrogen embrittlement. ISO 11114-4:2017 International organization for standardization
2. ISO16573 (2015) Steel—measurement method for the evaluation of hydrogen embrittlement resistance of high strength steels. ISO 16573:2015 international organization for standardization
3. ISO17081 (2014) Method of measurement of hydrogen permeation and determination of hydrogen uptake and transport in metals by an electrochemical technique. ISO 17081:2014 International organization for standardization
4. Giuliani L, Mirabile M, Sarracino M (1974) Embrittlement kinetics of N 80 steel in H_2S environment. Metall Mater Trans B 5(9):2069–2073
5. Nagumo M, Yoshida H, Shimomura Y, Kadokura T (2001) Ductile crack growth resistance in hydrogen-charged steels. Mater Trans 42(1):132–137
6. Koyama M, Tasan CC, Tsuzaki K (2019) Overview of metastability and compositional complexity effects for hydrogen-resistant iron alloys: inverse austenite stability effects. Eng Fract Mech 214:123–133
7. Troiano AR (2016) The role of hydrogen and other interstitials in the mechanical behavior of metals. Metallogr Microstruct Anal 5(6):557–569
8. Peral LB, Zafra A, Fernández-Pariente I et al (2020) Effect of internal hydrogen on the tensile properties of different CrMo(V) steel grades: influence of vanadium addition on hydrogen trapping and diffusion. Int J Hydrog Energy 45(41):22054–22079
9. Venezuela J, Hill T, Zhou Q et al (2021) Hydrogen-induced fast fracture in notched 1500 and 1700 MPa class automotive martensitic advanced high-strength steel. Corr Sci, 109550
10. Duportal M, Oudriss A, Savall C et al (2022) On the implication of mobile hydrogen content on the surface reactivity of an austenitic stainless steel. Electrochim Acta, 139684
11. Alekseeva E, Belyaev A, Zegzhda A et al (2018) Boundary layer influence on the distribution of hydrogen concentrations during hydrogen-induced cracking test of steels, diagnostics, resource and mechanics of. Mater Struct 3:43–57
12. Polyanskiy VA, Belyaev AK, Alekseeva EL et al (2019) Phenomenon of skin effect in metals due to hydrogen absorption. Contin Mech Thermodyn 31(6):1961–1975
13. Martinsson A, Sandström R (2012) Hydrogen depth profile in phosphorus-doped, oxygen-free copper after cathodic charging. J Mater Sci 47(19):6768–6776
14. Wu R, Ahiström J, Magnusson H et al (2015) Charging, degassing and distribution of hydrogen in cast iron (report SKB R-13-45), Svensk Kärnbränslehantering AB. Swedish Nuclear and Waste Management Co., Stockholm
15. Oriani RA (1972) A mechanistic theory of hydrogen embrittlement of steels. Berichte der Bundesgesellschaft für physikalische Chemie. 76(8):848–857
16. Troiano AR (1960) The role of hydrogen and other interstitials in the mechanical behavior of metals. Trans ASM 52:54–80
17. Birnbaum HK, Sofronis P (1994) Hydrogen-enhanced localized plasticity—a mechanism for hydrogen-related fracture. Mater Sci Eng A 176(1–2):191–202
18. Sofronis P, Birnbaum HK (1995) Mechanics of the hydrogen dislocation-impurity interactions. I. Increasing shear modulus. J Mech Phys Solid 43(1):49–90
19. Taha A, Sofronis P (2001) A micromechanics approach to the study of hydrogen transport and embrittlement. Eng Fract Mech 68(6):803–837
20. von Fick A (1855) Ueber diffusion. Ann Phys 170(1):59–86

21. Shewmon PG (1963) Diffusion in solids. Mc Graw-Hill Book Co Inc., New York
22. Kolachev BA (1985) Vodorodnaya khrupkost metallov [Hydrogen embrittlement of metals]. Metallurgiya, Moscow (in Russian)
23. Gorsky W (1935) Theorie der elastischen Nachwirkung in ungeordneten Mischkristallen (elastische Nachwirkung zweiter Art), Physikalische Zeitschrift der Sowjetunion 8:457–471
24. Serebrinsky S, Carter EA, Ortiz M (2004) A quantum-mechanically informed continuum model of hydrogen embrittlement. J Mech Phys Solids 52(10):2403–2430
25. McLean D (1957) Grain boundaries in metals. Clarendon Press, London
26. Irwin GR (1957) Analysis of stresses and strains near the end of a crack traversing a plate. J Appl Mech 24(3):361–364
27. Irwin GR (1958) Fracture. In book: Encyclopedia of Physics, Ed. by Flugge S, Vol. 6: Elasticity and Plasticity, Springer, Berlin (1958) 551–590
28. Wilkins ML (1999) Computer simulation of dynamic phenomena. Springer Science & Business Media, Berlin, Heidelberg
29. Srawley JE (1976) Wide range stress intensity factor expressions for ASTM E 399 standard fracture toughness specimens. Int J Fract 12:475–476
30. Bessonov NM, Golovashchenko SF, Volpert VA (2009) Numerical modelling of contact elastic-plastic flows. Math Model Nat Phenom 4(1):44–87
31. Wilkins ML (1964) Calculation of elastic-plastic flow. In: Alder B (Ed), Methods in computational physics. 3, pp 211–263.
32. Hirth JP (1980) Effects of hydrogen on the properties of iron and steel. Metallurg Trans A 11(6):861–890
33. Herms EE, Olive JM, Puiggali M (1999) Hydrogen embrittlement of 316L type stainless steel. Mater Sci Eng A 272(2):279–283
34. Li X, Feng Zh, Song X, Wang Y, Zhang Y (2022) Effect of hydrogen charging time on hydrogen embrittlement of CoCrFeMnNi high-entropy alloy. Corrosion science 198, 110073

Chapter 16
Investigation of Single-Phase Flow Characteristics in Roughness Microchannels Using 3D Boundary Element Method

Olga Solnyshkina, Aiguzel Bulatova, Nazgul Fatkullina, and Iskandar Garifullin

Abstract To predict the behavior of reservoir fluids in the porous medium and study them at the macro level, a detailed study of hydrodynamic flows in the porous medium at the microscale at the level of individual pores, taking into account their structural features. This work is dedicated to the application of the computational approach to study the single-phase flow characteristics in roughness microchannels. Numerical approach is based on the accelerated tree-dimensional Boundary Element Method. The influence of irregularities of different shapes and their relative arrangement on the side walls of microchannels on the viscous fluid flow in the Stokes regime has been studied. The results of this work can also be useful in the design of microfluidic devices, as well as microheat exchangers.

Keywords Microchannel · Roughness · Flow patterns · Stokes flow · Numerical simulation · Boundary element method · Fast multipole method

16.1 Introduction

Nowadays, technologies based on the application of microfluidic systems consisting of microchannels of various configurations and their networks are widely used in various fields of industry, since this approach allows miniaturization, integration and automation of complex flow processes by reducing volume, as well as providing compactness and precise control of flow characteristics. Computer simulation of flows in complex three-dimensional areas is relevant to microhydrodynamics in the construction of labs-on-a-chip applied, for example, to the study of multi-stage biochemical processes in which one of the control mechanisms is mixing of

O. Solnyshkina (✉) · A. Bulatova · N. Fatkullina · I. Garifullin
Center for Micro- and nanoscale dynamics of dispersed systems, Ufa University of Science and Technology, 450076 Ufa, Russia
e-mail: olgasolnyshkina@gmail.com

H. Altenbach and V. Kuzkin (eds.), *Advanced Problems in Mechanics*,
https://doi.org/10.1007/978-3-031-80103-7_16

reagents. The manufacturing of microdevices poses a number of other urgent problems. For example, the design of efficient micro heat exchangers, which can find application in microelectronics. Multiparametric study of fluid flow characteristic in microchannels is of importance for device design and for implementation of correct control of micro mechanical and microfluidic devices. In addition, simulation of fluid filtration in reservoir micromodels is of practical importance for development of enhanced oil recovery technologies, since macro-effects arising during oil filtration in the reservoir are largely determined by the features of fluid dynamics in pore channels. Using such approaches reduces the number of laboratory and field experiments, optimizes existing production mechanisms and strategies, and assesses potential risks. Since pore channels have natural irregularities of various scales, inevitably emerging features on their surfaces and shape deviations from the idealized cylindrical or plane-parallel shape considered in most works must be taken into account and thoroughly analyzed. One of the generally accepted ideologies of representation of the porous medium is the capillary model, when the volume of the porous medium is considered as a network of microchannels of complex shapes. A number of experimental works have revealed a significant discrepancy between the measured integral characteristics of flows in porous media and the assumed theoretical results. Therefore, adequate three-dimensional numerical modeling of such processes is extremely important.

Nowadays, the influence of wall roughness on the hydrodynamic flow patterns is a subject of active study. In [1] it was shown that depending on the type of fluid flow, the effects of the walls can vary. It is considered that the larger the cross-sectional area through which fluid filtration occurs, the less influence of the channel walls on it, besides the influence of Van der Waals forces and electrokinetic forces are not so important as in the case of a smaller channel cross-sectional area. In [2] the results of numerical simulation of viscous fluid dynamics in a microchannel with rough walls are presented. As the relative roughness of the walls increased, a sharp deviation of the streamlines and the vortex formation were observed.

Some articles also present simulation of viscous fluid flow in microchannels of different configuration with various shapes (rectangle, trapezoid) obstacles along the walls. In experimental work [3], the water flow through a microchannel with roughnesses in the form of trapezium, uniformly spaced along the length of the channel is considered, the change of flow rate and pressure drop along the length of the channel was studied. As a result of comparing the obtained data with the classical theory of laminar flow, some discrepancy was found. Therefore, adequate three-dimensional numerical simulation of such processes to more deeply understand the effect of surface roughness on microfluidic flows is extremely important. In [4], the three-dimensional case considered the flow between two plates, the surface of one of which was uniformly covered by rectangular obstacles of equal height. The study showed a significa nt influence of the obstacle geometry on the flow characteristics. In [5] similar processes, but for a serpentine microchannel were numerically studied in the three-dimensional case. The surface was covered by triangular, rectangular or trapezoidal obstacles. There is also an approach in which the near-wall layer with roughnesses is averaged and treated as an element of a porous medium in which flow is described by the Darcy equations [6]. In [7], the irregularities or roughnesses of the channel are described as an averaged homogeneous porous layer characterized by porosity and height. Fluid flow in such a layer is described by Darcy's law, i.e. no direct study of the flow structure around the roughness is performed.

Among the numerical methods of three-dimensional modeling of continuum mechanics problems, the most popular are finite-difference methods [8], finite element method, control volume method [9] and boundary element method (BEM) [10], which differ between themselves in the way the calculation area is represented and mathematical basis. For modeling single-phase and multiphase flows in regions with complex geometry in the three-dimensional case, the BEM is effective. Direct calculations by the boundary element method of Stokes flow of single-phase liquids and emulsions for the most part have been conducted for channels of symmetric shapes or in the two-dimensional case [11, 12]. One reason for this is related to the fact that to solve the three-dimensional asymmetric problem, even for relatively simple shapes, the computational grid on the channel surface contains tens of thousands of boundary elements. At present, new methods based on algorithmic and hardware acceleration of calculations are able to calculate such flows. For example, BEM can be accelerated using fast multipole method (FMM) [13, 14] on heterogeneous computing architectures. It is this approach that is used to solve the problems in this paper.

The purpose of this work is to apply the above approach to study the features of hydrodynamic flows in microchannels with irregularities of different shapes and locations on the side walls of the channel.

16.2 Problem Formulation and Numerical Approach Description

In the frame of this work we consider the periodic flow of a viscous incompressible fluid in a channel of a complex asymmetric shape driven by a constant pressure drop. The flow at low Reynolds numbers are typical for microhydrodynamics. In such type of regimes, the viscosity forces are much more significant than inertial forces associated with acceleration or deceleration of fluid particles, which makes it possible to completely neglect inertial terms in the calculations. Thus, fluid flow can be described by Stokes equations

$$-\nabla p + \mu \nabla^2 \mathbf{u} = \mathbf{0}, \quad \nabla \cdot \mathbf{u} = 0, \tag{16.1}$$

where $\mathbf{u}$ is the velocity, μ is the dynamic viscosity, and p is the pressure.

The surface of the simulated channel S is considered as a set of side surface S_{side}, inlet S_{in} and outlet S_{out} sections: $S = S_{in} \cup S_{out} \cup S_{side}$. It is assumed that the channel is infinite, the surface S_{side} and the flow are repeated with a period L in the direction of the axis x. On the surface of the channel side walls S_r the non-slip condition is set, which is completed by the flow periodicity conditions

$$\mathbf{u}|_{x=0} = \mathbf{u}|_{x=L} = \mathbf{u}_s, \quad \mathbf{f}|_{x=L} = -\,\mathbf{f}|_{x=0} + \mathbf{f}_p = \mathbf{f}_s, \quad \mathbf{f}_p = \mathbf{i}_x \Delta p, \tag{16.2}$$

where Δp is the given pressure drop on the considered channel section L, $\mathbf{u}_s$ and $\mathbf{f}_s$ are the unknown functions on the inlet and outlet cross-sections of the channel, $\mathbf{f}$ is the traction.

To solve the described problem (16.1)–(16.2), the approach developed by the authors based on the boundary element method is applied. The boundary integral formulation for the case of a fluid with viscosity μ, occupying a volume V, bounded by the surface S with normals $\mathbf{n}(\mathbf{x})$, directed outside the fluid volume [10], is

$$\begin{aligned}
\mathbf{u}(\mathbf{y}) - \int_S \mathbf{K}(\mathbf{y},\mathbf{x}) \cdot \mathbf{u}(\mathbf{x})\, dS(\mathbf{x}) &= \frac{1}{\mu}\int_S \mathbf{G}(\mathbf{y},\mathbf{x}) \cdot \mathbf{f}(\mathbf{x})\, dS(\mathbf{x}), \quad \mathbf{y} \in V,\\
\frac{1}{2}\mathbf{u}(\mathbf{y}) - \int_S \mathbf{K}(\mathbf{y},\mathbf{x}) \cdot \mathbf{u}(\mathbf{x})\, dS(\mathbf{x}) &= \frac{1}{\mu}\int_S \mathbf{G}(\mathbf{y},\mathbf{x}) \cdot \mathbf{f}(\mathbf{x})\, dS(\mathbf{x}), \quad \mathbf{y} \in S,\\
-\int_S \mathbf{K}(\mathbf{y},\mathbf{x}) \cdot \mathbf{u}(\mathbf{x})\, dS(\mathbf{x}) &= \frac{1}{\mu}\int_S \mathbf{G}(\mathbf{y},\mathbf{x}) \cdot \mathbf{f}(\mathbf{x})\, dS(\mathbf{x}), \quad \mathbf{y} \notin S, V,
\end{aligned} \tag{16.3}$$

where $\mathbf{y}$ is the radius vector of considered point, $\mathbf{x}$ is the radius vector of the point on the surface S, $\mathbf{K}(\mathbf{y}, \mathbf{x}) = \mathbf{T}(\mathbf{y}, \mathbf{x}) \cdot \mathbf{n}(\mathbf{x})$, $\mathbf{G}$ and $\mathbf{T}$ are the second and third rank tensors, in the Cartesian coordinate system have the form

$$G_{ij}(\mathbf{y},\mathbf{x}) = \frac{1}{8\pi}\left(\frac{\delta_{ij}}{r} + \frac{r_i r_j}{r^3}\right), \quad T_{ijk}(\mathbf{y},\mathbf{x}) = -\frac{3}{4\pi}\frac{r_i r_j r_k}{r^5}, \tag{16.4}$$

$$r_i = y_i - x_i, \quad i, j, k = 1, 2, 3,$$

where δ_{ij} is the Kronecker symbol.

The surface of the channel is covered by a triangular mesh. Using the method of collocations in the centres of triangular elements, the equation with respect to the calculation points located on the channel surface is reduced to a system of linear algebraic equations (SLAE) with a dense asymmetric matrix. For correct calculation of flows in channels of complex shapes, their surface should be covered by a sufficiently detailed mesh with tens of thousands of triangular elements. In this case, there is unavoidably a lack of memory of the computational system, which constitutes the main computational difficulty in solving the obtained system. The described problem can be solved using the approach involving both algorithmic acceleration and hardware acceleration. The direct method for solving SLAE has been replaced by the iterative generalized minimal residual method [15], in which the matrix-vector multiplication module is implemented using the fast multipole method. This method was developed in the late 1980s to solve the N-body gravity problem and allows one to significantly reduce the computational complexity of the matrix-vector product from quadratic to linear. FMM is a highly parallelizable algorithm that allows for additional acceleration through the use of heterogeneous computational architectures and clusters [16]. In this paper we utilized an approach in which part of modules has been parallelized on CPU using OpenMP and part has been implemented on GPUs using CUDA technology. The approach was developed by the authors for the Stokes equations and was successfully applied to simulate a large number of emulsion droplets [13]. It was shown that the algorithm has linear complexity and that calling a single FMM for an SLAE with $5.3 \cdot 10^6$ unknowns took 7 seconds without taking into account the time taken to generate the data structure. But in contrast to the problem of droplet dynamics in an unbounded domain, the problem of the flow in channels of arbitrary shapes also arises with a poorly conditioned matrix of the system. In this paper, a modification of the generalized minimum residual method is used, in which a non-fixed matrix is used as the right preconditioner. A reduced precision FMM is employed for this purpose, while the global iterations employ a higher precision FMM. This approach was pioneered for solving the Stokes equations in [17]. The implementation of the fast multipole method is described in more detail in the paper [18].

16.3 Roughness Effect on Flow Characteristics

First of all, the program modules were validated by comparison of the calculated velocity values with the analytical solution for the viscous fluid flow in a channel of rectangular cross section with plane walls. The calculations were performed for Reynolds number $Re = 0.15$, $Re = \rho U_{max} d_h/\mu$, $d_h = 4HW/2(H + W)$ is the hydrodynamic diameter, H is the channel height, W is the channel width. When comparing the velocities in the xz plane, the relative error was 1.36%, and 2% in the yz plane, which is quite acceptable for BEM.

In addition, the numerical approach was tested for the case of droplet dynamics in a cylindrical channel [19]. A good agreement of droplet shapes with experimental results was shown for several values of internal and external fluid viscosity ratios and several ratios of characteristic droplet and channel sizes.

16.3.1 Considered Geometries

In this work, a qualitative triangulation of six variants of microchannel configurations was developed to study the effect of irregularities of different shapes and relative arrangement on the side walls of microchannels on hydrodynamic flows during viscous fluid flow. It should be noted that the computational time depends significantly on the quality of the mesh on the channel surface. Two types of irregularities were considered—in the form of squares and equilateral triangles. Geometric parameters and schematic representation of the channels are shown in Table 16.1 and on Fig. 16.1, where W is width of the narrow part of the channel, W_r is width of bump,

Table 16.1 Geometric parameters of microchannels with irregularities

	Roughness shape	L (W)	H (W)	W_r (W)	L_r	L_{sh}	N_Δ
case 1	⊓	3	0.35	0.5	0.5	0	58544
case 2	⊓	3	0.35	0.5	0.5	$0.5L_r$	44138
case 3	⊓	3	0.35	0.5	0.5	L_r	44252
case 4	∧	3	0.35	0.5	0.5	0	47468
case 5	∧	3	0.35	0.5	0.5	$0.5L_r$	47182
case 6	∧	3	0.35	0.5	0.5	L_r	47368

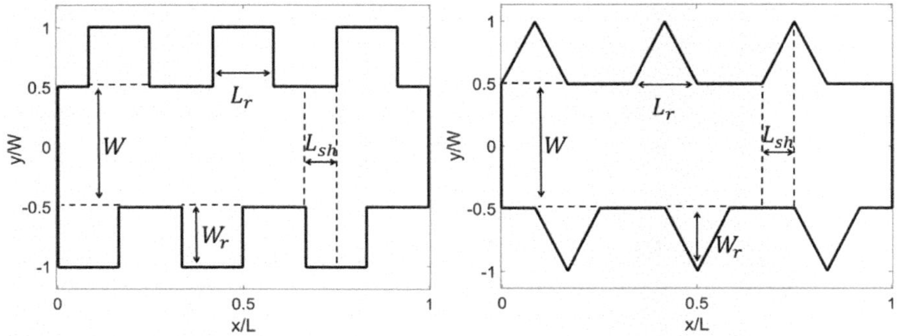

Fig. 16.1 Geometry of the microchannels with designed roughnesses

L_r is length of bump, L_{sh} is length of shift of irregularities relatively each other on x axis, N_Δ is the number of triangles of computational mesh. Calculations of high computational complexity in the designed channels were carried out at $Re = 0.15$.

16.3.2 Results of Single-Phase Flow Simulation

As a result of the calculations, the flow patterns for all variants of configurations of microchannels in the $z = H/2$ plane have been obtained. Visualization of the longitudinal and transverse velocity components is shown on Fig. 16.2 for channels with square roughnesses and on Fig. 16.3 for channels with triangular roughnesses. It is shown that the flow structure is affected not only by the shape of the roughnesses, but also by their location relative to each other (the presence of shift L_{sh}). Thus, in the case of irregularities located opposite each other along the length of the channel, there is a significant decrease in the maximum flow velocity in the area of irregularities. As the shear value increases, the velocity distribution changes, and as a result, for the configuration with bumps located with a shift relative to each other $L_{sh} = L_r$,

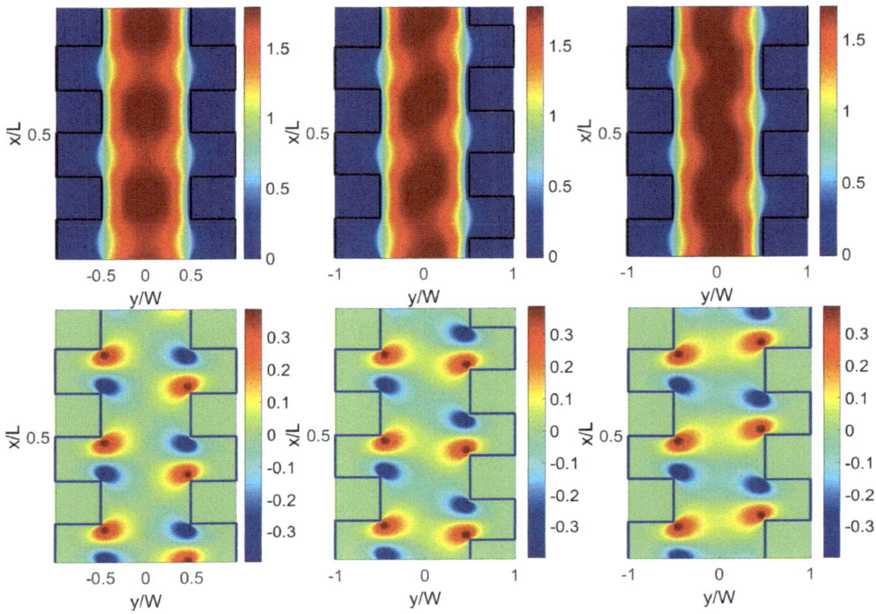

Fig. 16.2 Longitudinal $u_x/U_{average}$ (top row) and transverse $u_y/U_{average}$ (bottom row) velocity components in the plane $z = H/2$, $Re = 0.15$ for channels with square irregularities: case1, case2, and case 3 (from left to the right)

there is a more uniform velocity distribution with a smaller difference between the maximum and minimum values of the flow velocity.

For more detailed analysis of the influence of traps shape, velocity profiles inside channels in plane $z = H/2$ at different distance from the side wall relative to the channel centerline were calculated: $y = 0$, $y = 0.125\,W$, $y = 0.25\,W$, $y = 0.375\,W$. Figure 16.4 shows the results of calculating the velocity modulus relative to the average flow velocity at the inlet section of the channel in comparison with the velocity in a smooth channel of rectangular section (denoted in the figures by variant plane) for three different variants of bump shift: $L_{sh} = 0$, $L_{sh} = 0.5L_r$, $L_{sh} = L_r$. It can be seen that for the three distances from the channel wall, the velocity values are lower than those at the same distance for a smooth channel. At the same time, at the maximum approach to the wall in the case of a straight channel, the velocity becomes lower than the average velocity in rough channels. For all values of shift when moving closer to the channel wall, the behavior of the curves changes significantly, and two characteristic minima are distinguished—in the middle of the trap and in the middle of the gap between the traps. The influence of the trap geometry in the considered configurations is not so evident. The fluid flow velocities in the smooth channel section coincides practically. Some divergence is observed only in the area of irregularities. For channels with square roughness, velocity values are in average 2% less for all considered cases than for triangular roughnesses.

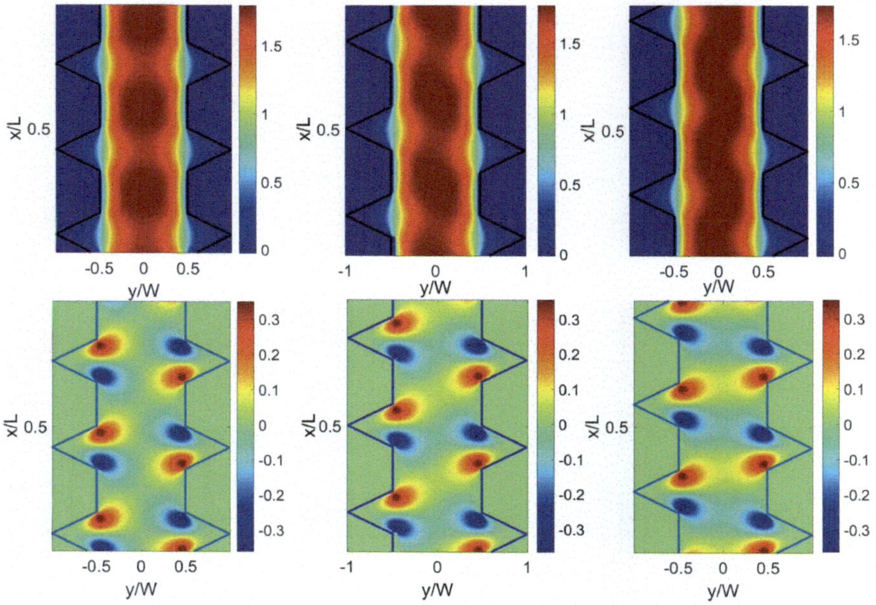

Fig. 16.3 Longitudinal $u_x/U_{average}$ (top row) and transverse $u_y/U_{average}$ (bottom row) velocity components in the plane $z = H/2$, $Re = 0.15$ for channels with triangular irregularities: case1, case2, and case 3 (from left to the right)

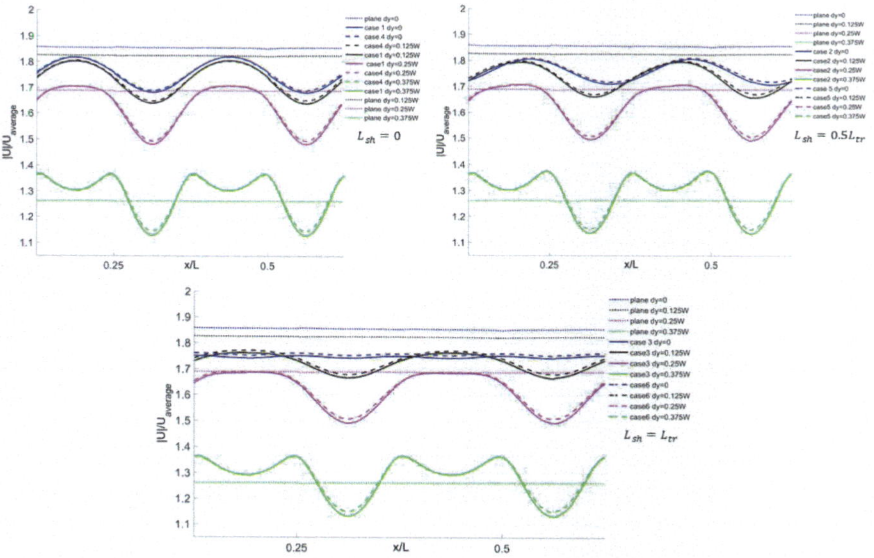

Fig. 16.4 Relative velocity profile of the fluid at different distances from the centerline in the microchannels in the plane $z = H/2$

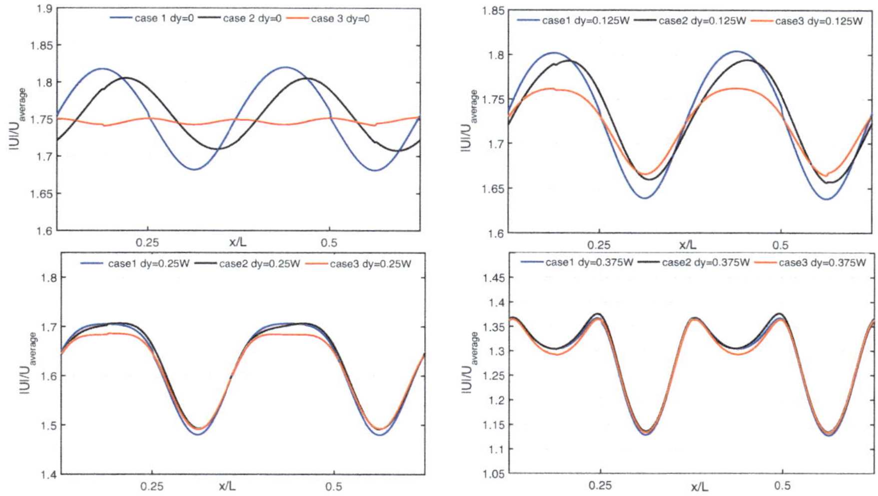

Fig. 16.5 Relative velocity profile of the fluid at different distances from the centerline in the microchannel with square roughness in the plane $z = H/2$

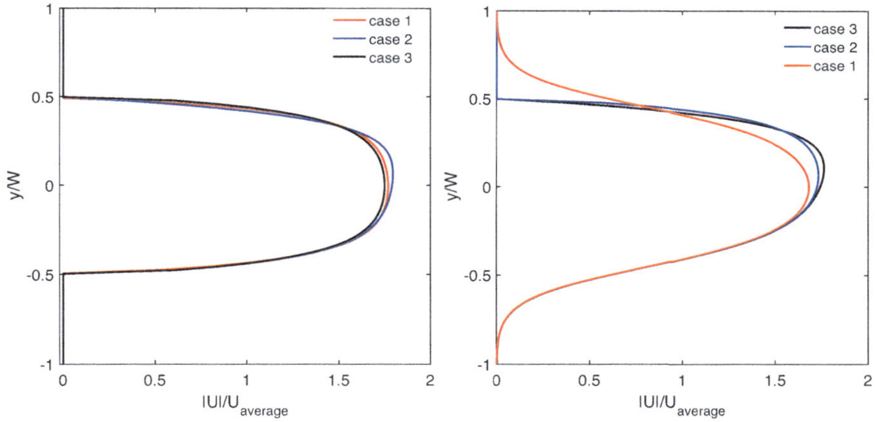

Fig. 16.6 Relative fluid velocity profile in the microchannel with square roughness in the plane $z = H/2$

For a more detailed consideration of the effect of trap shear relative to each other, Figs. 16.5 and 16.6 present relative velocity profiles in different planes for channels with square irregularities. Figure 16.5 shows that the change in velocity values along the length of the channel is not uniform for all cases. The curves at the centerline of the channels differ significantly for all values of the trap shear. When moving closer to the side wall, the difference in fluid velocity for channels with different shear values becomes minimal, that is, the influence of the second wall with irregularities is reduced to a minimum for the chosen ratio of the width of the narrow part of the channel to the trap parameters. Figure 16.6 shows the relative velocity profile in the channel cross section in the narrow part (left) and in the middle of the trap (right). For the bottleneck for the case $L_{sh} = 0.5L_r$ the velocity profile becomes asymmetric. The difference between the maximum achievable velocity values in this case does not exceed 1.5%. The maximum values in the middle of the trap differ more significantly. In the case of shift $L_{sh} = L_r$ the difference is 5%. It is possible that the width of the straight part of the channel should be smaller compared to the parameters of the roughness in order to have a more significant effect on the flow characteristics of the roughness.

16.4 Conclusions

As part of the work, a three-dimensional simulation of Stokes flows in microchannels with irregularities of different shapes and relative spatial distribution was carried out. Despite the fact that the formulation of the corresponding mathematical problems is carried out within the framework of the classical equations of mechanics and physics, it is necessary to involve modern computational methods and tools for their solution. We applied previously developed efficient codes, based on the BEM accelerated both due to a highly efficient scalable algorithm (Fast Multipole Method) and due to the use of heterogeneous computing architectures (multi-core CPUs and graphics processors). Single-phase flows in microchannels with irregularities have been performed. Flow patterns were obtained for all configurations of microchannels in several cross-section planes. It is shown that the flow structure is influenced not only by the form of irregularities, but by their location relative to each other. The proposed approach and the obtained results can be used in research of practically significant problems associated with the single and multiphase flows in porous media. Furthermore, the conducting numerical experiments to identify the features of viscous fluid flows in microchannels with roughness, which can be regarded as a set of obstacles and cavities of different shapes, sizes and locations is important for the determination of parameters to control the dynamics of dispersed particles in the viscous fluid flow by using the appropriate geometric features of microfluidic devices.

Acknowledgements The reported study was funded by the Russian Science Foundation within the research project No.21-79-10212.

References

1. Farhan Shafique M, Laister A, Clark M, Miles RE, Robertson ID (2011) Fabrication of embedded microfluidic channels in low temperature co-fired ceramic technology using laser machining and progressive lamination. J Europ Ceramic Soc 31(13):2199–2204
2. Li Z, Wan J, Zhan H, He L, Huang K (2020) An energy perspective of pore scale simulation and experimental evidence of fluid flow in a rough conduit. J Hydrol 587:125010. https://doi.org/10.1016/j.jhydrol.2020.125010
3. Weilin Q, Mala GhM, Li D (2000) Pressure-driven water flows in trapezoidal silicon microchannels. Int J Heat Mass Transfer 43:353–364
4. Hu Y, Werner C, Li D (2003) Influence of three-dimensional roughness on pressure-driven flow through microchannels. J Fluids Eng 125(5):871–879
5. Rawool AS, Mitra SK, Kandlikar SG (2006) Numerical simulation of flow through microchannels with designed roughness. Microfluidics and Nanofluidics 2:215–221
6. Kleinstreuer C, Koo J (2004) Computational analysis of wall roughness effects for liquid flow in micro-conduits. J Fluids Eng 126(1):1–9
7. Judy J, Maynes D, Web BW (2002) Characterization of frictional pressure drop for liquid flows through microchannels. Int J Heat Mass Transfer 45(17):3477–3489
8. Kovaleva LA, Musin AA, Zinnatullin RR, Akhatov ISh (2011) Destruction of water-in-oil emulsions in electromagnetic fields. ASME 2011 international mechanical engineering congress and exposition, IMECE 2011. V 6, pp 617–621. https://doi.org/10.1115/imece2011-62935
9. Valiullina VI, Zamula YuS, Mullayanov AI, Iulmukhametova RR, Musin AA, Kovaleva LA (2022) Experimental and numerical study of the water-in-oil emulsion thermal motion in rectangular cavity with a heated bottom. Advanced problem in mechanics II, pp 117–125. Springer, Cham. https://doi.org/10.1007/978-3-030-92144-6_9
10. Pozrikidis C (1992) Boundary integral and singularity methods for linearized viscous flow. Cambridge University Press, Cambridge, MA
11. Roca JF, Carvalho MS (2013) Flow of a drop through a constricted microcapillary. Comput Fluids 87:50–56. https://doi.org/10.1016/j.compfluid.2012.11.020
12. Yin B, Luo H (2013) Numerical simulation of drops inside an asymmetric microchannel with protrusions. Comput Fluids 82:14–28. https://doi.org/10.1016/j.compfluid.2013.05.005
13. Abramova OA, Pityuk YA, Gumerov NA, Akhatov IS (2014) An efficient method for simulation of the dynamics of a large number of deformable droplets in the stokes regime. Doklady Phys 59(5):236–240. https://doi.org/10.1134/S102833581405005X
14. Itkulova YA, Solnyshkina OA, Gumerov NA (2013) Boundary element simulations of compressible bubble dynamics in stokes flows. IMECE2013-63200, V07BT08A010
15. Saad Y (2000) Iterative methods for sparse linear system, 2nd edn. SIAM, Philadelphia
16. Gumerov NA, Duraiswami R (2008) Fast multipole methods on graphics processors. J Comput Phys 227(18):8290–8313
17. Itkulova YA, Solnyshkina OA, Gumerov NA (2012) Toward large scale simulations of emulsion flows in microchannels using fast multipole and graphics processor accelerated boundary element method. IMECE2012-86238, pp 873–881. https://doi.org/10.1115/IMECE2012-86238
18. Abramova OA, Pityuk YA, Gumerov NA, Akhatov IS (2019) Three-dimensional simulation of Stokes flow around a rigid structure using FMM/GPU accelerated BEM. Commun Comput Inform Sci 965:427–438. https://doi.org/10.1007/978-3-030-05807-4_36
19. Solnyshkina OA, Fatkullina NB, Bulatova AZ (2020) Three-dimensional simulation of drop motion in channels of different cross-sections. J Phys Conf Ser 1675(1):012099. https://doi.org/10.1088/1742-6596/1675/1/012099

Chapter 17
Modelling of Water Huff and Puff Technology in a Low Permeability Oilfield

Oleg I. Tsykunov

Abstract This paper presents a study of the applicability of the water Huff and Puff technology in tight oil reservoirs using a dual porosity hydrodynamic model. This technology uses capillary imbibition to extract oil from the tight pores of a hydrophilic reservoir. To simulate this process, a synthetic hydrodynamic model was created in tNavigator software with model of dual porosity and permeability. Small pores were analogous to the matrix and large pores were analogous to cracks. Since spontaneous capillary imbibition for the dual media model is not provided in this package, a script was created to convert the additional pressure gradient between media caused by capillary pressure into a change in matrix permeability. The simulation results show an increase in oil production after the operation. Pilot works were carried out at 4 wells. Two of them were indicative. In one well there was an increase in oil production. Comparison of the oil production dynamics before and after water Huff and Puff with the results of hydrodynamic modeling shows good convergence even with a synthetic model without adaptation to a specific well.

17.1 Introduction

The exploration of oilfields with hard-to-recover reserves is becoming an actual problem. The main category of such fields is tight reservoirs. The main difficulty in its exploration is low efficiency of the reservoir pressure maintenance system due to very low permeability (less than 1 mD), which leads to a low oil recovery factor and low well flow rates. There are a large number of methods for solving this problem–injection of various gases into the well, hydraulic fracturing, thermal methods and others. But the cheapest and most affordable is the periodic water injection. This technology is called water Huff and Puff and it is unique because there is only pilot works of this technology in the USA and in PRC.

O. I. Tsykunov (✉)
Peter the Great St. Petersburg Polytechnic University, St. Petersburg 195251, Russia
e-mail: olegtsykunovmgdn@gmail.com

H. Altenbach and V. Kuzkin (eds.), *Advanced Problems in Mechanics*,
https://doi.org/10.1007/978-3-031-80103-7_17

17.2 Water Huff and Puff

17.2.1 Technology

Water Huff and Puff is cyclical technology. Each cycle consists of three stages. At the first stage, the oil production well is stopped and water is pumped into it. At the second stage, the well is stopped for a certain time (from several days to a month). At the last stage, the well returns to production. The cycle repeats as long as the effect of additional oil production exceeds the cost of well transfer.

This technology is very cheap since no compounds and reagents are pumped into the well and the main operation is re-piping the well into injection.

17.2.2 Physical Principle

The physical principle of this technology is to initiate spontaneous capillary imbibition during water injection. If the reservoir is hydrophilic water under the influence of the capillary pressure will soak into small pores and displace oil into large pores [1]. This process is described by the Laplace equation, which links the contact angle (θ), , surface tension (σ), , and capillary radius (r_c) (1). The smaller the radius is, the higher the pressure is and, consequently, the technology can be more effective in tight reservoirs.

$$P_k = \frac{2\sigma cos\theta}{r_c} \tag{17.1}$$

According to the histogram of the distribution of pore channels in the core of a low-permeability field and their participation in filtration (Fig. 17.1), it can be seen that more than half of the pores (less than 0.57 microns) are not involved in filtration. If the well is operating with zero water cut, then these pores are not covered by exploration. If free water appears in the reservoir (for example, after injection), then the water will soak into these small pores and displace oil from them into the system of large pores [6].

Since the reservoir pressure maintenance system is often inefficient in low-permeability fields, many wells operate with minimally profitable oil rates with zero water cut. It is proposed to use the technology on such wells.

17.2.3 Brief Overview of Application Experience

Water huff and puff technology has been used in the USA and China.

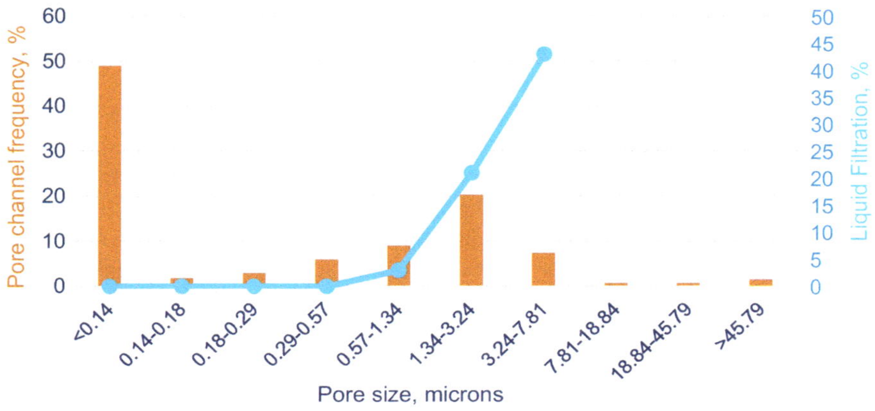

Fig. 17.1 Histogram of the distribution of the frequency of pore channels and their participation in filtration on the core of a low-permeability field

Pilot works in the United States were carried out in the Bakken shale oil reservoirs in North Dakota and the Parshall field [2, 7]. Since the reservoir is hydrophobic, then, based on the Laplace equation, the technology should not give additional oil production. This is exactly what happened, that's why they currently use a variation of the huff and puff technology with carbon dioxide injection, which does not use capillary forces, but affects the viscosity of the oil.

The application experience in China is successful. More than 40 different experiments were carried out on horizontal wells with multi-stage hydraulic fracturing [3–5, 8, 9]. More than 95% of the operations were successful and gave a significant increase in oil production [8]. So, for example, pilot work at the Yangchang field on the hydrophilic reservoir Chang 7, the average oil production rate before the operation is 9.9 tons/day, and after one cycle–18.2 tons/day. The average incremental oil production per well for all pilot operations in China is 972 tons [8].

Since the geological and physical characteristics of the Chinese reservoirs are similar to the low permeability Russian oil fields it is expected that the use of water huff and puff technology will also be successful.

17.3 Hydrodynamic Modeling of the Efficiency of a Water Huff and Puff

17.3.1 3D Hydrodynamic Model

A synthetic hydrodynamic model was created in the tNavigator software to analyze the efficiency.

To speed up the calculation of the model, it was decided to model only a quarter of the reservoir, and place the well at the origin. Thus, all flow rates must be multiplied by 4 to obtain the flow rates of a real well.

Filtration has been reduced to two-dimensional since vertical flows do not affect the process. The reservoir was created in the form of a 500 × 500 m parallelepiped with a thickness of 10 m. Hydraulic fracture was created. The size of the x and y cells is 5 m. At the same time, a local grid refinement of the grid up to 0.5 m was created in the fracture area of the hydraulic fracturing to reproduce boundary effects. Analysis of the sensitivity of the model to the cell size showed that the difference between the cell sizes of 10 × 10 m and 20 × 20 m is less than 1%. At the same time, without local grinding of the mesh, the difference increases to 5%, which is why it is necessary to use it.

17.3.2 Capillary Imbibition Model

The process of capillary imbibition was set using the dual porosity model. Since there is a system of large pores through which filtration takes place and a system of small pores in which there is no filtration. It can be compared with the matrix-crack system where the crack is large pores, the matrix is a system of small pores. For modeling fractured formations, there is a function of dual porosity, in which the formation is defined by two layers that are superimposed on each other. The matrix is given by a medium with a horizontal permeability of 0.045 mD, and the permeability of the fracture system is 0.45 mD. In this case, the porosity was distributed between the media according to the real proportions of pores of each medium in the rock sample.

The conductivity coefficient between the media is also set, it defined as:

$$Tr = C_{darcy} K V \sigma \tag{17.2}$$

where K is the permeability of the rock, V is the volume of the matrix in the cell, σ is the sigma factor, C_{darcy} is the Darcy constant.

Since the simulator cannot specify an additional pressure gradient between the matrix and the fracture the only way out is to convert this gradient into additional permeability. However, the permeability cannot be increased by a multiplier equal to the additional gradient since it must be multiplied by the ratio of the sum of drawdowns to the initial drawdown without imbibition. Then there is the problem of determining the depression between environments.

To solve this problem, it is necessary to determine the sigma factor. This coefficient physically expresses the contact area between the fracture and the matrix. Calculated according to the formula:

$$\sigma = 4\left(\frac{1}{l_x^2} + \frac{1}{l_y^2} + \frac{1}{l_z^2}\right) \tag{17.3}$$

where l_x, l_y, l_z are block sizes in directions.

There is no data about the size of the blocks since the two media are evenly distributed in each other, therefore, the sigma factor should tend to infinity. But this will lead to infinite conductivity between the media and in real filtration the matrix is not involved in the development and does not exchange phases with the crack. But if we set the sigma factor to zero, then the exchange between the phases will be impossible. Therefore, it is necessary to introduce it in such a way that the depression funnel in the matrix medium lags behind the fracture medium in the model. As a result, the sigma factor was taken equal to 0.1. This value is a free parameter for this model and it was selected iteratively. With a lower value than 0.1, mass transfer between media with different filtration and capacitive properties becomes almost impossible, with a higher filtration rate tends to infinity and we will not see the effect of capillary imbibition.

Precalculations were run to determine the drawdown between medias as a consequence of the lag of the drawdown funnel in the two environments. On average, this value ranged from 0.9–1.2. Since in the SCHEDULE section and in tNavigator arithmetic it is impossible to refer to the pressure difference between the cells (it is possible to work with only one specific cell in each step), the drawdown was set to 1 bar.

Based on the research on capillarometry on the core, the average value of the capillary pressure in the medium of small pores was taken to be 20 bar. Since the capillary pressure will be an additional pressure gradient between the fracture medium and the matrix it must be specified in the model.

The equation for fluid filtration in a porous medium is described by the Darcy equation:

$$Q = \frac{k}{\mu}\frac{\Delta P}{L}F \tag{17.4}$$

where Q is the flow rate, m^3/s; k – permeability, m^2; μ – viscosity, Pa•s; L—length, m; F is the cross-sectional area of the channel, m^2; ΔP – pressure difference, Pa.

Since the value of the additional pressure drop caused by capillary forces and denoted below as ΔP_{cap} is known, it can be substituted into the Darcy filtration Eq. (17.4).

$$Q = \frac{k}{\mu}\frac{\left(\Delta P + \Delta P_{cap}\right)}{L}F \tag{17.5}$$

In the analytical calculation, it remains only to substitute all known values and obtain the oil flow rate from the matrix to the fractured medium but since the calculation is performed in a hydrodynamic simulator it is proposed to recalculate the increase in pressure drop into an increase in the proportionality coefficient - permeability.

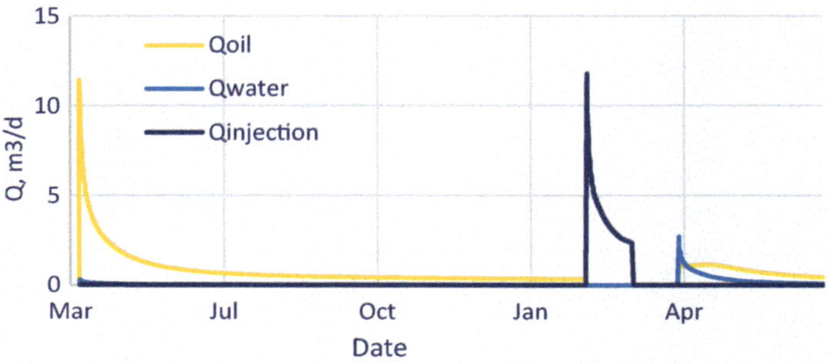

Fig. 17.2 Graph of oil production and water injection for one huff and puff cycle

$$\frac{\Delta P + \Delta P_{cap}}{\Delta P} \cdot k = k_{cap} \tag{17.6}$$

where k_{cap} is modified permeability, m^2.

This logic was embedded in the script using the ARITHMETHIC keyword–at each calculation step the condition of the water saturation at which the imbibition starts is checked and then if there is an excess the permeability of the matrix is multiplied by the calculated coefficient depending on capillary pressure, formation pressure in the cell, initial permeability.

The results of the calculation of one cycle are shown in the Fig. 17.2. At the beginning there is a normal work of a production well. Then water is injected and the well is stopped for imbibition. Then production begins. First, the water cut is high, then it falls and the oil production rate increases. Additional oil production is 30 tons in one cycle, since in the model there is a quarter of the reservoir, then in a real well the increase is 120 tons.

17.4 Pilot Works and Comparison with the Model

Pilot work was carried out at four wells. The tests were carried out on waterless marginal wells, which were recently transferred to the reservoir pressure maintenance stock. Stopping them and turning them into production represents one cycle of Huff and Puff. Three directional wells and one horizontal well were selected.

At the first well, it was not possible to meet the required injection and shutdown times, which were significantly higher than the optimal time. The second well had problems with reservoir pressure after injection. It was 1.5 times less than before injection. This makes wells 1 and 2 unrepresentative. In wells 3 and 4 injection and shutdown times were optimal and these wells are indicative. In well 3 oil flow rate did not change after the operation. Excess oil production was recorded in well 4. The Fig. 17.3 shows the daily flow rate of well 4.

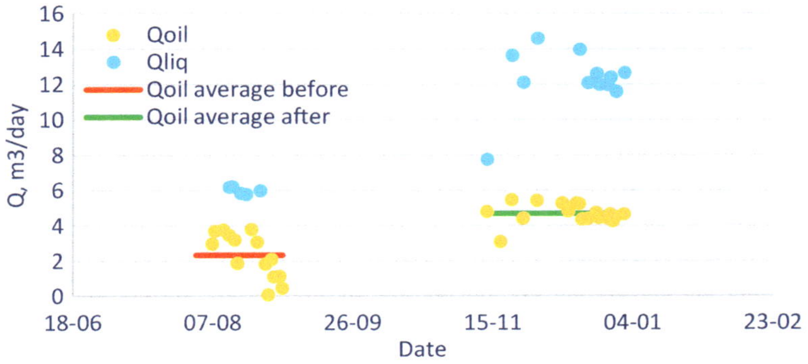

Fig. 17.3 Daily flow rates of well 4 before and after huff and puff

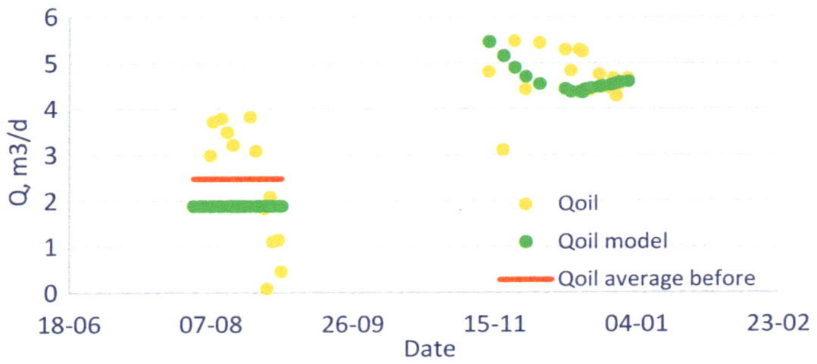

Fig. 17.4 Comparison of model and pilot works for well 4

As can be seen from the well 4 production schedule, the oil rate before the operation was unstable, however, the average oil rate was 2.3 m^3/day. The flow rate after one Huff and Puff cycle increased to an average of 4.7 m^3/day.

If we compare the results of pilot work with the results of hydrodynamic modeling for a single well with similar injection and shutdown periods (one month each) shown in Fig. 17.4 we can conclude that the model converges with the fact.

The average flow rate before the operation according to the model is 1.9 m3/day, and after - 4.6 m3/day. The discrepancy with the actual oil production rate before the operation is 23%. This is due to the fact that a synthetic model of an average well was used without adaptation to the history of work, moreover, in absolute values, the discrepancy is only 0.6 m3/day. The discrepancy with the oil production rate after the operation is only 2%.

17.5 Conclusions

Based on the results of this work, the following conclusions can be drawn:

1. A hydrodynamic model of a single well and a model of capillary impregnation were created. To simulate the process of capillary imbibition it was required to create a model of dual porosity and permeability. To simulate the pressure gradient between small and large pore media it was necessary to introduce an additional conversion function of the capillary pressure into an additional pressure drop depending on the water saturation of the cell. The results of the model calculation showed that under ideal conditions the technology is effective
2. Pilot work has shown that there is a positive effect from the use of water huff and puff technology. Of the 4 wells, only two were representative. In well 3, the oil flow rate did not change, which indicates the operation of the capillary impregnation effect, since the flow rate did not fall with a decrease in the oil phase permeability. In well 4, there was an increase in oil production from 2.3 to 4.7 m3/day.
3. Comparison of the results of pilot work with the model showed good convergence, despite the fact that the hydrodynamic model was synthetic. The discrepancy in oil production after the operation was 2%, which indicates that the proposed technique for modeling capillary impregnation can be used to design injection using this technology.

References

1. Chen T, Yang Z, Ding Y, Luo Y, Qi D, Lin W, Zhao X (2018) Waterflooding huff-n-puff in tight oil cores using online nuclear magnetic resonance. Energies 11:1524. https://doi.org/10.3390/en11061524
2. Sheng JJ (2017) Critical review of field EOR projects in shale and tight reservoirs. J Pet Sci Eng 159:654–665. ISSN 0920–4105. https://doi.org/10.1016/j.petrol.2017.09.022
3. Li X-H (2015) Huff-n-puff water injection in tight Duha field. Spec Oil GAS Reserv 22(4):144–146
4. Lin Y-Y, Wang P-P, Li Q-D, Han B-M, Li H-L, Wu Y-P, He W-S (2016) Performance analysis of different modes of Huff-n-puff water injection in horizontal wells in An-83 Chang-7 tight oil reservoir. Petrochem Ind Appl 35(6):94–97
5. Qin G, Dai X, Wu M, Zheng Y, Qing Y, Li W (2020) Study of massive water huff-n-puff technology in tight oil field and its field application. J Petrol Sci Eng. https://doi.org/10.1016/j.petrol.2020.107514
6. Teklu TW, X Li, Zhou Z, Alharthy N, Wang L, Abass H (2018) Low-salinity water and surfactants for hydraulic fracturing and EOR of shales. J Pet Sci Eng 162:367–377. https://doi.org/10.1016/j.petrol.2017.12.057

7. Todd HB, Evans JG (2016) Improved oil recovery IOR pilot projects in the bakken formation Paper presented at the SPE low perm symposium, Denver, Colorado, USA
8. Wang X, Xie K, Zhang J, Zhang Y, Zhang Y, Wang W, Yan X, Zhao (2021) Study on the key influential factors on water Huff-n-Puff in ultralow-permeability reservoir. Geofluids 2021:Article ID 5885366, 8 pages. https://doi.org/10.1155/2021/5885366
9. Z Li, X Qu, W Liu, Q Lei, H Sun, Y He (2015) Development modes of Triassic Yanchang Formation Chang 7 Member tight oil in Ordos Basin. NW China. Pet Explor Dev 42(2):241–246

Chapter 18
Analysis of the Equilibrium of a Magnetic Contactless Suspension

Pavel Udalov, Alexey Lukin, Ivan Popov, and Dmitriy Skubov

Abstract In this paper, we consider a one-dimensional model of magnetic suspension in the case of a variable magnetic field. Analytical expressions for the averaged over a period of an external current source for the suspension equilibrium position are obtained. It is shown that the magnetic stiffness of the suspension has a sign-variable character in the general case. An expression for the averaged value of the magnetic stiffness is obtained and it is shown that this value is always positive.

Keywords Levitation · Inertial sensor · Contactless suspension · Micro-electro-mechanical system

18.1 Introduction

A non-contact magnetic suspension (see Fig. 18.1) in the simplest case is a system consisting of two basic elements—an inductor coil, to which an alternating current is applied, and a proof levitating mass (PM) made of a conductive material. Due to the law of electromagnetic induction, currents are induced in the PM when current flows in the coil, and an electromagnetic force arises between the parts of the structure. Due to this effect it is possible to use the PM, for example, as a stand or support for engineering and other structures.

The main purpose of this work is an analytical study of the magnetic part of the electromagnetic suspension and obtaining conditions for finding the equilibrium

P. Udalov (✉) · A. Lukin · I. Popov · D. Skubov
Peter the Great St. Petersburg Polytechnical University, Saint Petersburg, Russia
e-mail: pp_udalov@mail.ru

A. Lukin
e-mail: lukin_av@spbstu.ru

I. Popov
e-mail: popov_ia@spbstu.ru

D. Skubov
IPME RAS, Saint-Petersburg, Russia

H. Altenbach and V. Kuzkin (eds.), *Advanced Problems in Mechanics*,
https://doi.org/10.1007/978-3-031-80103-7_18

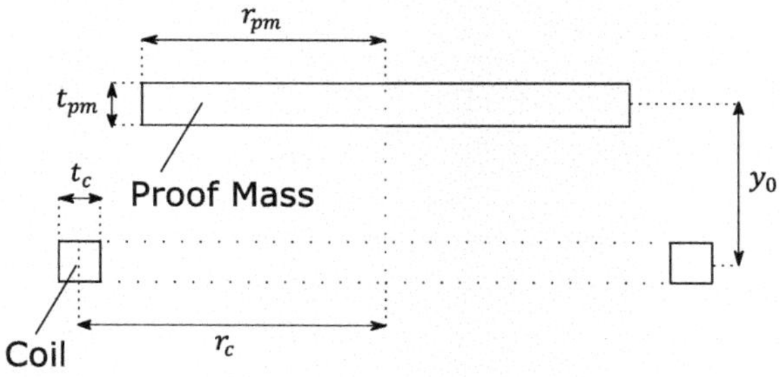

Fig. 18.1 Scheme of magnetic suspension

position and stability of vibrations of a levitating object near its equilibrium. The magnetic stiffness of the magnetic suspension is also evaluated. It is shown that in this formulation, it has a harmonic character, which should be considered when adjusting the electrical part of the system to eliminate the magnetic stiffness of the system.

18.2 Mathematical Model

The energy of the magnetic field W of the coil of radius r_c and thickness t_c powered by alternating current i_1 and the disc-shaped proof mass (PM) of the thickness t_{pm}, radius r_{pm} with induced eddy current i_2 is

$$W = \frac{1}{2}L_1 \dot{\iota}_1^2 + M_{12}(y)\dot{\iota}_1\dot{\iota}_2 + \frac{1}{2}L_2\dot{\iota}_2^2, \quad (18.1)$$

where L_1 and L_2 are the self-inductance of coil and PM, respectively [1].

$$L_1 = \mu_0 r_c w^2\left(\ln\frac{8r_c}{0.44705t_c} - 2\right), \quad L_2 = \mu_0 r_{pm}\left(ln\frac{16r_{pm}}{0.44705t_{pm}} - 2\right), \quad (18.2)$$

where w is the number of turns of coil, $M_{12}(y)$ is the mutual inductance, μ_0 is the magnetic permeability of free space, $\dot{()} = \frac{d}{dt}$, t is the time, y is the distance between coil and PM.

In general, the mutual inductance $M_{12}(y)$ is a complex non-analytic function. This represents the main difficulty for analytical investigation of the magnetic suspension model [2]. However, it is possible to take into account some features of the micro-machine suspension device, which allow the application of the mutual inductance approximation formula [3]. These simplifications consist in the assumption that the linear dimensions of the coil and control mass are much larger than the height of the levitation equilibrium position y_0. It is also assumed that the induced eddy current i_2 is distributed along the levitated control mass such that a contour corresponding to the maximum eddy current density can be identified [2].

The eddy current contour is geometrically defined as a circle having the same diameter as the reference mass [2]. Due to the above features of the device, the force interaction in the vertical direction is reduced to the interaction between the eddy current and the levitating coil current [4]. If we consider the reference mass and the induced current as circles, the mutual inductance between the reference mass and the eddy current can be described by the Maxwell formula [3] as follows:

$$M_{12}(\kappa) = \mu_0 w\sqrt{r_c r_{pm}}\left[\left(\frac{2}{\kappa} - \kappa\right)K\left(\kappa^2\right) - \frac{2}{\kappa}E\left(\kappa^2\right)\right],$$
$$\kappa^2(y) = \frac{4r_c r_{pm}}{(r_c + r_m)^2 + y^2}, \tag{18.3}$$

where $K(\kappa)$ and $E(\kappa)$ are complete eliptic integrals of the first and second kind, $\kappa(y)$ is the elliptic modulus.

In deriving the equations of dynamics of PM, we consider the dynamics of its center of gravity. From this assumptions potential Π and kinetic T energies of the PM are

$$\Pi = \tilde{m}gy, \quad T = \frac{1}{2}\tilde{m}\dot{y}^2, \tag{18.4}$$

where $\tilde{m}$ is the mass of the PM, g is the gravity acceleration.

The dissipation function of the system Ψ can be written as follow

$$\Psi = \frac{1}{2}\mathrm{R}_2 \imath_2^2 + \frac{1}{2}\mu\dot{y}^2, \tag{18.5}$$

where R_2 is PM's electrical resistance, μ is the coefficient of mechanical friction.

We assume that the current i_1 generated in the coil by the current generator has the following form

$$\imath_1 = \imath_\mathrm{a}\sin\omega t, \tag{18.6}$$

where i_a and ω are the amplitude and the high frequency of the current i_1, correspondingly.

We use the Lagrange-Maxwell equations to write equation of motion of the upper ring. The displacement y and the current i_2 are taken as generalized coordinates:

$$\frac{d}{dt}\frac{\partial W}{\partial \imath_2} + \frac{\partial \Psi}{\partial \imath_2} = 0,$$
$$\frac{d}{dt}\frac{\partial T}{\partial \dot{y}} + \frac{\partial(\Pi - T)}{\partial y} + \frac{d\Psi}{d\dot{y}} = 0. \tag{18.7}$$

Substituting equations (18.2–18.6) into equation (18.7), we obtain

$$L_2 i_2 + \frac{dM_{12}(y)}{dy}\dot{y}\imath_a\sin\omega t + M_{12}(y)\imath_a\omega\cos\omega t + R_2\imath_2 = 0,$$
$$\tilde{m}\ddot{y} + \mu\dot{y} - \frac{dM_{12}(y)}{dy}\imath_2\imath_a\ \sin\omega t + \tilde{m}g = 0. \tag{18.8}$$

We introduce the nondimensional quantities as [5]

$$\xi = \frac{y}{2r_c}, a = \frac{r_{pm}}{r_c}, \beta = \frac{\mu_0 r_c}{L_1}, \gamma = \frac{t_c}{t_{pm}}, j_1 = \frac{i_1}{i_a}\ j_2 = \frac{i_2}{i_a}, r = \frac{R_2}{L_1\omega},$$
$$m_{12} = \frac{M_{12}}{L_1}, l = \frac{L_2}{L_1}, \tau = \omega t, \varepsilon = \frac{g}{2\omega^2 r_c}, \alpha = \frac{L_1 i_a^2}{2\tilde{m} g r_c}, \lambda = \frac{\mu}{\tilde{m}\omega}. \tag{18.9}$$

and rewrite equation (18.8) in nondimensional form as

$$lj_2' + rj_2 = -\frac{dm_{12}(\xi)}{d\xi}\xi' sin\tau - m_{12}(\xi)cos\tau,$$
$$\xi'' + \lambda\xi' = \varepsilon\left(\alpha\frac{dm_{12}(\xi)}{d\xi}j_2 sin\tau - 1\right), \tag{18.10}$$

where the prime indicates the derivative with respect to τ. The small nondimensional quantity $\varepsilon\alpha = \frac{L_1 i_a^2}{4r_c^2\omega^2}$ defines the ratio between magnetic and electrical energies.

For further study of the system (18.10) we find its equilibrium position $(j_2, \xi) = (j_{20}, \xi_0)$, where $j_{20} = j_{20}(\tau)$; ξ_0 is the constant.

18.3 Equilibrium

We assume the condition under which the PM reaches an average value $\xi_0 = \frac{y_0}{2r_c}$ (y_0 is the dimensional average equilibrium point) over the period of change in coil current i_1. In other words, due to the harmonicity of the excitation current i_1, the magnetic force acting on the PM is also harmonic, which leads to a steady-state oscillatory process of the PM relative to some average value ξ_0. When the magnetic force changes, the equilibrium point changes, which corresponds to the equilibrium of the gravity force and the Ampere force also acting on the PM in the considered formulation. Moreover, it is worth noting that, in the general case, a variable magnetic field forms a variable magnetic stiffness, as will be shown later.

$$lj_{20}' + rj_{20} = -\frac{m_{12}(\xi_0)}{2}e^{i\tau} + c.c,$$
$$\frac{i\alpha}{2}\frac{dm_{12}(\xi_0)}{d\xi}j_{20}e^{i\tau} + \frac{1}{2} + c.c = 0. \tag{18.11}$$

where $c.c$ is the complex conjugate value of equation [6].

Solving first equation in (18.11), we obtain the steady-state soultion for the current j_{20} as

$$j_{20} = \frac{im_{12}(\xi_0)}{2\sqrt{r^2 + l^2}}e^{i(\tau+\phi)} + c.c, \tag{18.12}$$

where $cos\phi = \frac{l}{\sqrt{(l^2+r^2)}}$, $sin\phi = \frac{r}{\sqrt{(l^2+r^2)}}$, $i^2 = -1.$, $i^2 = -1$.

Substituting equation (18.12) in the second equation in (18.11), we integrate the current j_{20} within $\tau \in [0, 2\pi]$ and obtain the expression for the average equilibrium position ξ_0.

$$m_{12}(\xi_0)\frac{dm_{12}(\xi_0)}{d\xi} = -\frac{2(r^2 + l^2)}{\alpha l}, \tag{18.13}$$

Figures 18.2 and 18.3 show the dependence of the value of the equilibrium position ξ_0 on the physical parameters of the system.

Figures 18.2 and 18.3 show that with increasing parameter α (which corresponds to increasing current i_a the value ξ_0 increases, which is associated with an increase in electromagnetic force of the ring and, therefore, Ampere force. When the parameter r increases (which corresponds to a decrease in frequency ω), the equilibrium position ξ_0 takes smaller values, which is caused by an increase in reactance.

After finding the condition on the integral position of the suspension equilibrium, it is interesting to estimate the magnetic stiffness for its subsequent compensation by including an electric field, which is studied in the papers [2].

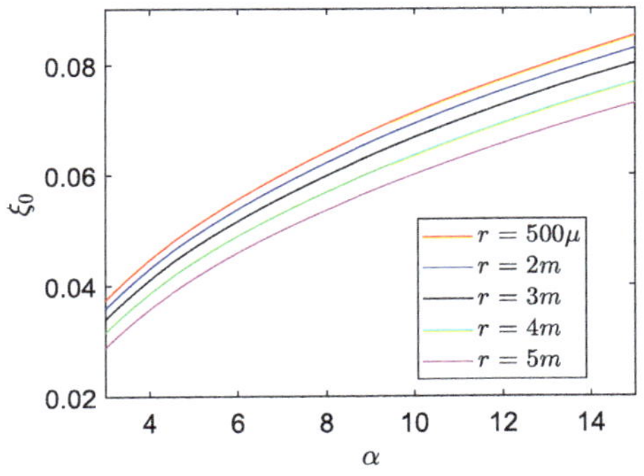

Fig. 18.2 Dependence of the integral position of the frame equilibrium ξ_0 on the parameter α with ($r = 500\mu$; $2m$; $4m$; $5m$)

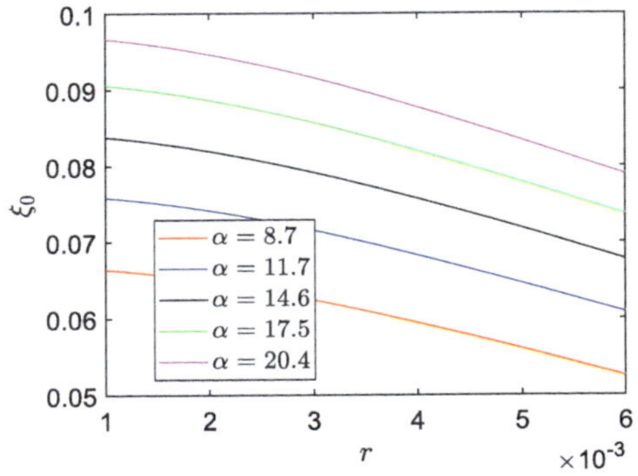

Fig. 18.3 Dependence of the integral equilibrium position ξ_0 on the parameter r ($\alpha = 8.7$; 11.7; 14.6; 17.5; 20.4)

18.4 Magnetic Spring Constant of the Suspension

Let us study the behavior of the PM near to the equilibrium point ξ_0. It is assumed that the linear displacement of the PM ξ is small in comparison with ξ_0, hence the following inequality

$$\frac{\xi}{\xi_0} \ll 1, \tag{18.14}$$

Because of (18.14), the function of the mutual inductance $m_{12}(\xi)$ can be extended by a Taylor series at the point ξ_0 in the form can be written as follow

$$m_{12}(\xi) = m_{12}(\xi_0) + m_{12}'(\xi_0)(\xi - \xi_0) + m_{12}''(\xi_0)(\xi - \xi_0)^2, \tag{18.15}$$

Substituting (18.15) into the last equation of set (18.10) and taking into account (18.13) the differential equation of the linear displacement of the PM near to the equilibrium point can be written as

$$\xi'' + \lambda\xi' + c_m\xi = F_\xi, \tag{18.16}$$

where F_ξ is generalized force acting on the PM, c_m is the spring constant of the magnetic suspension

$$c_m = \frac{\varepsilon\alpha\left[m_{12}m_{12}'' + \left(m_{12}'(\xi_0)\right)^2\right]}{\sqrt{r^2 + l^2}} \sin\tau \sin(\tau + \phi). \tag{18.17}$$

Equation (18.17) show that the magnetic stiffness has a periodic character and, in the general case, is sign-variable, which means an oscillatory mode of PM levitation near the integral equilibrium position. For a more detailed study of the dependence of the stiffness c_m on the system parameter, consider the average value of the stiffness $\langle c_m \rangle = \frac{1}{\pi}\int_0^\pi c_m(\tau)d\tau$

$$c_m = \frac{\varepsilon\alpha l\left[m_{12}m_{12}'' + \left(m_{12}'(\xi_0)\right)^2\right]}{2\left(r^2 + l^2\right)}. \tag{18.18}$$

The eqn (18.18) shows that the average value of the magnetic stiffness is always positive. The only possible problem is related to the sign-variability of the initial stiffness (18.17) which should be considered in the modes of operation associated with the compensation of magnetic stiffness by adding electrodes to the system and creating a negative electric stiffness, which is done, for example, in the paper [2].

18.5 Conclusion

In this paper, a magnetic suspension consisting of an inductance coil and a levitating disk was considered. When the coil is energized by alternating current, it is shown that in this case the character of the equilibrium position has an alternating harmonic form. Assuming that the equilibrium position comes to the same point during the coil current period, an estimate of the integral mean value of this quantity was given. An expression for the magnetic stiffness of the suspension is obtained and it is shown that it also has a periodic form, which should be taken into account in the construction of devices of this type.

Funding The research is funded by Russian Science Foundation grant № 21–71-10,009, https://rscf.ru/en/project/21-71-10009/.

References

1. Grover F (2004) Inductance calculations: working formulas and tables. Dover Publication, Mineola, NewYork
2. Poletkin KV, Shearwood CAIC (2012) Proposal for micromachined accelerometer, based on a contactless suspension with zero spring constant. IEEE Sens J 12(7):2407–2413
3. Rosa EB, Frederick WG (1948) Formulas and tables for the calculation of mutual and self-inductance. No. 169. US government printing office, Washington, D.C
4. Lu Z, Poletkin KV, Wallrabe U, Badilita V (2014) Performance characterization of micromachined inductive suspensions based on 3D wire-bonded microcoils. Micromachines 5:1469–1484
5. Dmitrii S, Ivan P, Udalov P (2021) Suspension of conductive microring and its gyroscopic stabilization, PREPRINT (Version 1) at Research Square. [https://doi.org/10.21203/rs.3.rs-171012/v1]
6. Nayfeh AH (2008) Perturbation methods. Wiley, New York

Chapter 19
Experimental Study of a Model Emulsion Stratification in a Heated from Above Rectangular Cell

Vilena I. Valiullina, A. R. Ayupov, A. A. Musin, and L. A. Kovaleva

Abstract One of the effective methods of emulsion separation is thermal. Heat leads to a large number of droplet collisions and to a decrease in the viscosity of the oil, which provides a higher chance of settling. Nevertheless, it is necessary to select the modes of thermal action on the emulsion system so that this does not lead to the opposite result. It was noticed as a result of the work that droplet deposition occurs more intensively in the central part of the cell and slows down in the wall areas, which is associated with the movement occurring in the emulsion, directed downwards in the center of the cell and up along the walls. The upward flow slows down the deposition of emulsion droplets in areas close to the walls.

19.1 Introduction

The emulsion is a complex multiphase system consisting of two or more immiscible liquids and emulsifiers [1]. There are many works aimed at studying the dynamics of motion of emulsion systems, both numerical [2, 3], and experimental [4]. The features of the dynamics of multiphase media are also experimentally studied using the microfluidic approach [5, 6].

According to [7], water-in-oil emulsions are of the greatest interest in oil production, because about 95% of the oil of this emulsion is produced in the world. Proper mixing or the presence of surfactants added to the volume of oil contributes to the formation of stable emulsions during oil production. Thus, it is necessary to apply methods for the separation of emulsions in order to reduce the problems associated with transportation at production facilities [8]. There are mechanical, chemical, electrical and thermal methods of stratification of dispersed systems. Mechanical methods include delamination under the influence of gravitational and centrifugal forces, chemical methods use various demulsifiers, also to achieve the delamination of the emulsion, it can be affected by an electric field [9].

V. I. Valiullina (✉) · A. R. Ayupov · A. A. Musin · L. A. Kovaleva
Ufa University of Science and Technology, Ufa, Russia
e-mail: valiullina.vilena@mail.ru

H. Altenbach and V. Kuzkin (eds.), *Advanced Problems in Mechanics*,
https://doi.org/10.1007/978-3-031-80103-7_19

One of the effective methods of emulsion separation is thermal. Heat plays an important role and has both advantages and disadvantages when used to destabilize emulsions. An increase in temperature affects the difference in viscosity and density of water and oil. In addition, this leads to a large number of droplet collisions and to a decrease in the viscosity of the oil, which reduces the strength of the accompanying stabilizing agents on the droplets and thus provides a higher chance of settling [10]. Consequently, a lower interfacial viscosity of the oil during shear will improve the diffusion rate of film drainage between adjacent droplets. Nevertheless, it is necessary to select the modes of thermal action on the emulsion system so that this does not lead to the opposite result [11].

The purpose of this present work is to experimentally investigate the behavior of emulsion systems under thermal action from above.

19.2 Experimental Research

19.2.1 The Object of the Study

The object of research in this work is a water-in-oil emulsion. Distilled water (Milli-Q) was used as the dispersed phase. The dispersion medium was vaseline oil. The surfactant SPAN 80 was used to obtain a stable emulsion. The method of preparing the emulsion was as follows: SPAN 80 and water were added to the vaseline oil, the resulting mixture was stirred at a rotation speed of 300 rpm using a top-drive agitator. The mass concentration of water in the emulsion was 5%. No separation of the emulsion into separate phases was observed during the entire experiment. Then microscopic studies of the size distribution of water droplets in the emulsion were carried out (see Fig. 19.1).

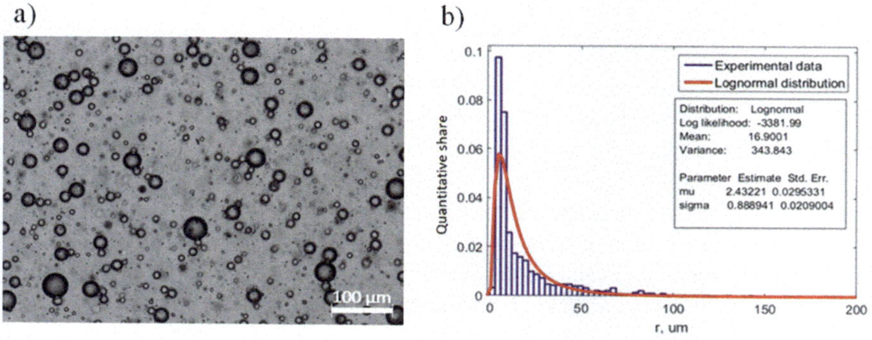

Fig. 19.1 Photo of emulsions under a microscope (**a**) and a histogram of the distribution of water droplets in the emulsion by size (**b**)

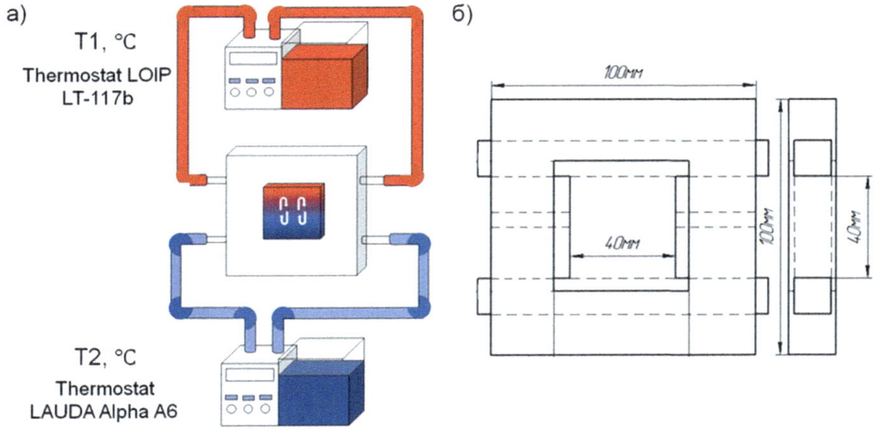

Fig. 19.2 Scheme of experimental setup (**a**) and convective cell (**b**)

The results showed that the average size of water droplets in the emulsion in the main volume was r ≈ 15 μm.

19.2.2 Experimental Setup and Methodology of the Experiment

Experimental studies of the process of precipitation of water droplets in an emulsion under thermal exposure were carried out in a rectangular cell made using 3D printing technology. Tubes were located in the upper and lower parts of the cell for the possibility of heating the faces using two thermostats LOIP LT-117b and LAUDA Alpha A6. The water temperature varied in the range of 23 ÷ 38°C in increments of 5°C. The deposition process was recorded on a camera. The scheme of the experimental setup is shown in Fig. 19.2.

An image of the emulsion system in the inner cavity of the cell was taken every 30 s during the experiment. The resulting images were processed in the ImageJ application software package.

19.2.3 Results of Experimental Studies

Three series of experiments were conducted. The lower face of the cell was maintained at a temperature of 23°C, the upper face was heated at 5, 10 and 15°C. Figure 19.3 shows characteristic photographs of the state of the emulsion system at

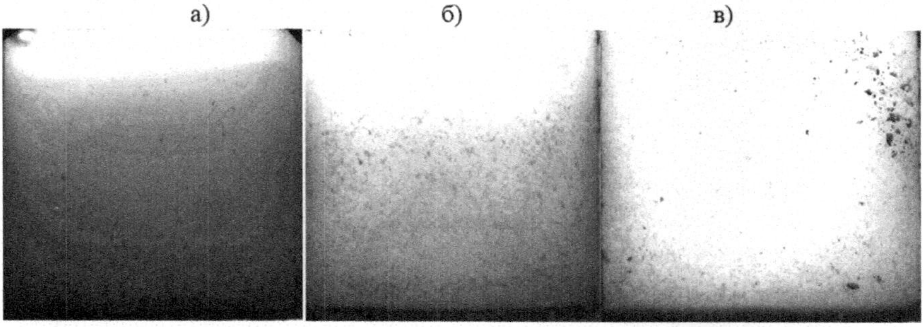

Fig. 19.3 Photos of the state of the emulsion system at a time of 60 min with a temperature difference: **a** ΔT = 5°C; **b** ΔT = 10°C; **c** ΔT = 15°C

time t = 60 min. The contrast of the image depends on the intensity of the transmitted light.

It can be seen from the figures that by the time of 60 min, an area with a high intensity of transmitted light (light area) is formed in the upper part of the cell, which indicates a decrease in the concentration of emulsion droplets in this area. This area is the larger the higher the temperature difference, which is associated with a decrease in the viscosity of the oil with an increase in temperature, and, accordingly, with a decrease in the resistance to droplet deposition. At the same time, it is seen that droplet deposition occurs more intensively in the central part of the cell and slows down in the wall areas. Analysis of the research results showed that when droplets are deposited in the emulsion system, movement is initiated, directed downwards in the center of the cell and up along the walls. The upward flow slows down the deposition of emulsion droplets in areas close to the wall. The higher the temperature drop in the system, the more noticeable this phenomenon is.

In the lower part of the cell, an area with a high degree of darkening is formed, which indicates the formation of a layer with a high concentration of emulsion droplets. The layer of settled droplets changes with an increase in the temperature drop in the cell. The results of digital processing of images obtained during three series of experiments are shown in Fig. 19.4 in the form of dependencies of the height of the exfoliated water layer over time.

As shown in Fig. 19.4, the droplet deposition rate increases with increasing temperature difference non-linearly. This is due to the presence of a nonlinear dependence of the viscosity of petroleum jelly oil on temperature. It can be seen that the difference in droplet deposition increases with time. This is due to the fact that at the initial moment of time, the temperature of the emulsion in the lower part of the cell is the same in all three cases and droplet deposition occurs at almost the same speeds (time interval from 0 to 30 min in Fig. 19.4). Over time, the high temperature zone in the upper part of the cell expands, which leads to an increase in the deposition rate drops (the time interval from 30 to 60 min in Fig. 19.4). In this time interval, precipitation of larger droplets is observed. Further, there is a slowdown in the growth of the layer of settled droplets (the time interval from 60 to 120 min in Fig. 19.4),

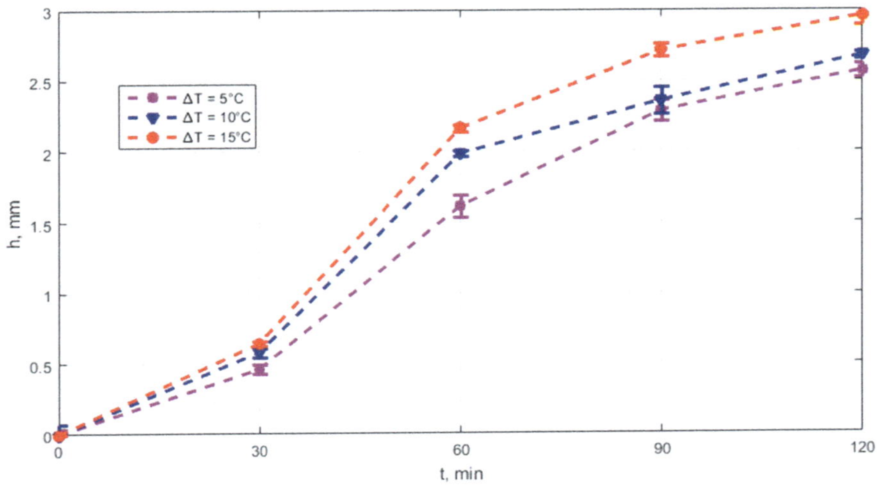

Fig. 19.4 The dependence of the height of the exfoliated water on time at various temperature differences (ΔT = 5°C, ΔT = 10°C, ΔT = 15°C)

which is associated with the deposition of smaller droplets, the rate of which is much lower. At the same time, the higher the temperature drop in the cell, the higher the layer of settled droplets.

19.3 Conclusions

The results of a study of the process of non-isothermal precipitation of droplets in a water-in-oil emulsion in a rectangular cell when heated from above and cooled from below are presented. A method has been developed for determining the distribution of the concentration of water droplets in the emulsion by the height of the cell based on the results of digital image processing, which allows more accurately determining the height of the formed layer of settled droplets. On the basis of the methodology, the dependences of the height of the exfoliated water on time at different temperature gradients are constructed and determined.

It was found that the degree of heating of the emulsion significantly affects the deposition time of water droplets. It has been observed that droplet deposition occurs more intensively in the central part of the cell and slows down in the wall areas, which is associated with the movement occurring in the emulsion system, directed downwards in the center of the cell and up along the walls. The upward flow slows down the deposition of emulsion droplets in areas close to the walls.

Acknowledgements The reported study was funded by the grant of the Russian Science Foundation (project №19-11-00298).

References

1. Mohyaldinn ME, Hassan AM, Ayoub MA (2019) Application of emulsions and microemulsions in enhanced oil recovery and well stimulation. Microemulsion-a chemical nanoreactor. IntechOpen
2. Pelusi F et al (2021) Rayleigh-Bénard convection of a model emulsion: anomalous heat-flux fluctuations and finite-size droplet effects. Soft Matter 17(13):3709–3721
3. Tukhbatova ER et al (2019) Investigation of thermohydrodynamic processes in oil-water emulsions under thermal influence. In: physical and mathematical modeling of processes in geomedium, pp 138–140
4. Ma F, Chen J, Zhang P (2018) Experimental study of the hydraulic and thermal performances of nano-sized phase change emulsion in horizontal mini-tubes. Energy 149:944–953
5. Solnyshkina OA, Batyrshin ES, Pityuk YuA (2021) Investigation of hydrodynamic flows in micromodels of double porosity media. Fluid Dyn 56(4):451–459
6. Abramova OA, Akhatov IS, Gumerov NA, Pityuk YA, Sametov SP (2018) Numerical and experimental study of bubble dynamics in contact with a solid surface. Fluid Dyn 53(3):337–346
7. Zolfaghari R et al (2016) Demulsification techniques of water-in-oil and oil-in-water emulsions in petroleum industry. Sep Purif Technol 170:377–407
8. Lee RF (1999) Agents which promote and stabilize water-in-oil emulsions. Spill Sci Technol Bull 5(2):117–126
9. Kovaleva L et al (2021) Influence of radio-frequency and microwave electromagnetic treatment on water-in-oil emulsion separation. Colloids Surf A PhysChemical Eng Asp 614:126081
10. Valiullina VI et al (2021) Experimental study of the stratification of polydisperse emulsions in a cell with heated walls. J Phys Conf Ser 2057(1):012041
11. Valiullina V et al (2022) Experimental and numerical study of the water-in-oil emulsion thermal motion in rectangular cavity with a heated Bottom. In: international summer school-conference advanced problems in mechanics, Springer, Cham, pp 117–125

Author Index

H. Altenbach and V. Kuzkin (eds.), *Advanced Problems in Mechanics*,
https://doi.org/10.1007/978-3-031-80103-7

The manufacturer's authorised representative in the EU is Springer
Nature Customer Service Centre GmbH, Europaplatz 3, 69115 Heidelberg,
Germany. If you have any concerns regarding our products, please
contact ProductSafety@springernature.com

Printed and bound by CPI Group (UK) Ltd, Croydon, CR0 4YY
01/10/2026
02225948-0001